Educational Challenges Regarding Military Action

Studies for Military Pedagogy, Military Science & Security Policy

Edited by H. Jung and W. Royl

Vol. 11

PETER LANG

Frankfurt am Main · Berlin · Bern · Bruxelles · New York · Oxford · Wien

Hubert Annen / Wolfgang Royl (eds.)

Educational Challenges Regarding Military Action

PETER LANG
Internationaler Verlag der Wissenschaften

Bibliographic Information published by the Deutsche Nationalbibliothek
The Deutsche Nationalbibliothek lists this publication in the Deutsche Nationalbibliografie; detailed bibliographic data is available in the internet at http://dnb.d-nb.de.

ISSN 1619-778X
ISBN 978-3-631-58941-0

www.peterlang.de

Contents

Editor's Preface

We are pleased to present to the interested professional public the 11th volume of the series "Studies for Military Pedagogy, Military Science & Security Policy". This international cooperation of military pedagogues owes its origin to the initiative of Edmund van Trotsenburg[1], who in 1989 organised the first conference of those peers willing to cooperate in this scientific discipline.

This occurred at a time when groups of the academic youth were inspired by the thought of laying the foundation stone for lasting piece through actions against military service. Nevertheless, the vision inherent to a world of ‚peace and freedom' has unfortunately not become reality. The development, including the current asymmetrical warfare, rather proved those right who advocated a defensive democracy and therefore saw their scientific duty in extending and deepening the aspect of military pedagogy within the education of a soldier. This regards on one hand the function and procedure of military training and education in general and on the other hand in particular the conveyance of military ethics as soldierly guiding principle.

Today, wars are no longer fought among nations only but between global-scale interest groups which do not feel bound to human rights and the law of armed conflicts. They are guided by fundamental ideologies or the most brutal form of profit seeking. Accordingly, the transfer of guidelines on professional ethics in particular for soldiers during peacekeeping operations is a conditio sine qua non. It is the only way to counter misconduct in regular troops as well.[2] Raising such a claim is one thing, putting it into action within practical military training remains the order of the day. This also applies to the members of the UN forces in Darfur said to exploit the hunger of the population by making girls and women sexually submissive through food.

The texts put together in volume 11 are revised contributions of authors who made a speech in 2006 in Bucharest (Romania) and 2007 in Jerusalem (Israel) within the scope of these two annual conferences of „International Association of

1 Prof. em. Dr. Edmund van Trotsenburg, Dean of Educational Science, University Klagenfurt (Austria)

2 Prunier, Gerard (2005): Darfur: The Ambiguous Genocide. New York: Cornell University Press.

Military Pedagogy“. The editors express their gratitude to Prof. Dr. Eng. Gabriel Malos, Defence University in Bucharest and Prof. Dr. Asa Kasher, Institute for Military Ethics, Defence University Jerusalem.

Hermann Jung Wolfgang Royl

Preface

The essays presented in here provide an international overview of military-pedagogical thinking and acting. In this, they reflect the sometimes close correspondence between the answers provided by military scholars to questions related to the content and function of military ethics and morale. These answers are so comprehensive as to suggest themselves as a starting point for further deliberations on military pedagogy but also in the fields of other applied pedagogic specialties.

The current requirement and existing demand for this is a consequence of armed conflicts, be they military or peacekeeping in nature, because the claim of military pedagogy – to translate ethical behaviour into effective action – takes up a moral demand that stems from the history of mankind, which is the tendency towards humanisation and democratisation. In this way, military pedagogy forms an intersection with those academic fields which in turn are the carriers and developers of societal and individual education.

The authors who have contributed towards this book make it clear, as a group, how the national defence of peace and freedom may be transformed into a pertinent international responsibility and competence for the safeguarding of world peace.

The essay by *Philip J. Rossi, SJ* has been placed at the beginning because it discusses the eternal question of whether wars can be just from the perspective of the Christian awareness of responsibility in this our century. The historical summary of the Just War Doctrine is followed by the moral analysis of consequences which result from both the nuclear and conventional worldwide armament potentials. Reasons of conscience to object against military service may thus be justified more comprehensively and intensively. (p. 17)

Asa Kasher is one of the foremost subject-matter experts on the difficult military situation in which the Israel Defense Forces (IDF) are opposing the Palestinian resistance forces. It is his concern to provide a rationale for the commitment of IDF troops in the Gaza Strip and to justify their deployment in moral terms. He does so in depth by referring to the principles of the just war theory and the topical question directed by the citizens of Israel to their government: "What have you done to protect me from a given violent threat that endangers my life?". (p. 25)

Juha Mäkinen discusses the importance of ethics and moral conscience for military training and education in general, and further develops thoughts that were central to the debates at the International Association of Military Pedagogy. This institution aims to create an acceptable framework for the training of soldiers into which military ethics and moral behaviour are embedded as principles. As for multinational troops committed in peacekeeping missions, this should result in a necessary, and thus binding, common base for moral behaviour both among themselves and when dealing with the civilian population. (p. 51)

Wolfgang Royl first calls to mind the philosophical tradition of the Occident. This is followed by an elaborate plea for ethical standards and moral modes of behaviour to be imparted in military initial and advanced training. This complements the reflections on training strategy given by Juha Mäkinen in his essay on peacekeeping operations. Guiding statements of Kant and Rawls are taken further by the high standards so vividly depicted by the Swiss theologian Hans Küng in his vision of the "world ethos". (p. 67)

Hermann Jung, in a directing position, has contributed significantly to the shaping of military pedagogy in the Austrian armed forces. Consequently, he advocates the linking of military intervention with on-site social stabilisation and cultural transformation in peacekeeping missions. To be up to the complex task, the operational forces require both general and mission-specific training on cultural competence. (p. 85)

Hubert Annen shows how military-pedagogical challenges need to be dealt with ever anew. Social changes in modern times require that one should recall to mind what is worth defending in Switzerland and the world. This justifies the efforts taken to ensure the army's operational readiness not only in terms of armament technology but also in mental terms. Leadership personnel have an important role in that. Thus, attention is to be accorded to the selection and training of military cadres. In the process, issues of military pedagogy – such as the proclaiming of values and the dealing with power and authority – need to be given a theoretical foundation and goal-oriented implementation. (p. 109)

Bo Talerud presents a view of practiced ethics in the military and Swedish national culture. What is imposing is how military training in the Scandinavian countries, and Sweden in particular, promotes the interest in soldiers as individuals. For Swedish national culture maintains an active relationship with a social heritage which goes back to the Danish pastor and adult educator Nikolaj Frederik Severin Grundtvig (1783-1872). It is on the work of N.F.S. Grundtvig, among others, that the high standards of democratic human education have been built, including in the military sphere. (p. 123)

Harold Simpson gives us an insight into the activities of a British advisory team. The team tackles the important task of supporting the armed forces of the Republic of Sierra Leone in developing and implementing a defence studies curriculum. Four years after the end of the civil war that raged there until 2001 it was possible to re-establish the conditions for a well-ordered self-defence of the country. The training goals to be implemented for that purpose were taxonomically structured so as to make possible a comparison with the scope and quality of international training standards. (p. 131)

Tamir Libel provides an interesting and instructive insight into the development of Israel's combat forces and the training of officer candidates. The title reveals a certain impatience on the author's part with the professionalisation efforts the Army necessarily had to undertake to successfully transform the former people's army. Given the threat situation created by the Intifada, the state of Israel had no alternative but to set the conditions for an effective military self-defence. (p. 145)

Juha Mäkinen's essay distinguishes itself in terms of military pedagogy by comparing civilian university training with the expectations on modern officer training in the light of the Bologna process. This training, his proposal reads, should be integrated into what is called a knowledge-creating process. When compared to conventional training, this brings out important new points, indeed: It places increased demands on the teaching staff which are met, among other aspects, by generally requiring of them a doctoral dissertation in a training-relevant field of science. This condition, in combination with their own relevant research studies, would serve to stabilise an important cooperative link between the civilian and military spheres. (p. 157)

Wolfgang Royl looks at the case of a German officer who refused to develop software for military purposes because it might have been used in the Iraq war. The officer maintained that the war was in violation of international law and, therefore, his mission was unconstitutional. The criteria on which the major's state of mind is analysed include three characteristics of the military profession: obedience, comradeship and conscience. In 2005, the judgment passed against him was rescinded by the highest German administrative court, as was his prior demotion. (p. 171)

Gavril Malos, Toma Plesanu & Sorina Mardar deal with the code of honour important to all armed forces, using the Romanian army's supply troops as an example and placing in the foreground those modes of behaviour that will benefit all and foster moral coherence among the troops. The values which need to be imbued in each soldier to reach that goal provide the ethical basis for reliability and mutual trust in everyday military activities. Ethical conduct is not always a matter of course. It depends upon established rules of behaviour and on soldiers setting mutual examples for themselves. (p. 191)

Radomir Saliger & Antonin Muller give information on how the Czech armed forces have succeeded in finding their place in the new security environment of the Western world. With the visible restructuring of the army, the conditions have been set for full membership in the North Atlantic Treaty Organisation. Informative illustrations depict the historical development of the military training system in the Czech Republic. This includes the structure and tasks of the Defence University in Vyškoc and the training requirements to be met by officer candidates. (p. 199)

Victoria Syme-Taylor considers thoroughness in the training of British field-grade officers to be a permanent military-pedagogical challenge for the teaching staff. What is important here, just as elsewhere in the alliance, is to avoid that the high standards of training are diluted through the routine of teaching. Students need to undergo further development of their personal performance abilities. Among other aspects (cf. the essay of David Hall (p. ...)), this is done by imparting comprehension of documents on historical combat and leadership achievements suitable for identification. (p. 209)

David Hall is to be commended for imparting to the reader an unexpected impression of the battlefields of France in the 19th and 20th centuries. Excursions to these places are conducted for British officers with the pedagogical intent in mind to develop their own leadership abilities by understanding tactical moves and comparing military operations on the ground. In the process, they are given the opportunity to prove their own strategic assessment skills. (p. 215)

Can Nakkas & Hubert Annen introduce the 'portfolio method', which is used to record learning results in an individual progression protocol. By recording the insights gained, which correspond with teaching and self-studies, a history log of individual learning is thus created that is traceable for each cadet. This enables both officer candidates and lecturers to use the portfolio text for the purpose of assessing whether learning goals and teaching contents have been met at the intended training level, and to what degree the conditions have been set for putting this into practice. (p. 227)

Antti-Tuomas Pulkka provides an informative insight into modern officer training in Finland. He shows how the instructors on site have accommodated the changes caused by the reorganisation of training into a university-type educational structure. It provides for, and calls for, students to study on their own initiative and responsibility. The developing sensitivity of the teaching staff, who are also the military superiors of the cadets, supports the latter's developing interests in research. (p. 235)

Luiza Kraft convincingly demonstrates how to succeed in imparting linguistic competence to students, using the Bucharest Defence University as an example. Cooperation within the Western alliance requires military personnel to be able to speak and write in English to communicate with each other. It has turned out, though, that foreign language teachers are put to additional tasks which go beyond imparting language skills. At times, they are required to explain military terminology and the cultural specifics of the partners within the alliance. (p. 243)

Hubert Annen, Zurich — Wolfgang Royl, Munich

1. Theoretical Considerations on Ethics and Morale

Philip J. Rossi, SJ

Historical and Cultural Contexts for "Just War" Doctrine

The "Disembedding" of the "Just War" Doctrine from the Contexts of its Origin and Development.

Many thinkers, in a line that extends back to the Stoics, goes through Augustine, Aquinas, Vitoria, Suarez, Grotius, and reaches down to present day expositors such as J. T. Johnson and Michael Walzer, have contributed to the establishment and refinement of the main concepts and argumentative strategies that, by the latter half of the twentieth century, gave the doctrine of "just war" a clearly recognizable formal shape. The just war doctrine, even as it has passed through the different conceptual matrices characteristic of each of these thinkers or schools of thought, seemed to have eventually emerged during the twentieth century as the exclusive intellectual property of no single one of them. One apparent result of this is that contemporary proponents of the just war doctrine need not articulate their acceptance and advocacy of the moral cogency of those principles and arguments in terms of an explicit commitment to one or another of the intellectual frameworks for moral thinking that historically shaped that doctrine. So even though there are contemporary proponents of just war who consider themselves Augustinians, others who consider themselves Thomists (though few, if any, are likely to consider themselves Suarezians!), there are also proponents of the just war doctrine who would not locate themselves intellectually as lineal inheritors of any of these founding traditions. In our contemporary context, one need not be a Stoic, an Augustinian, a Thomist, or a Suarezian, in order to advocate the moral relevance of the just war doctrine.

Of at least equal importance to this apparent dissociation of the just war doctrine from the intellectual frameworks – including some important religious and theological ones – in which it was first elaborated, is the recognition that one need not be a Stoic, an Augustinian, a Thomist, or a Suarezian, to *understand* the principles and arguments put forth, to engage them critically, or to dispute or refute them. The concepts and arguments of the just war doctrine have taken on a shape that gives them a formality that seems to render them intelligible – and, at least to some, morally cogent – apart from a broader commitment to the intellectual contexts in which they were either initially articulated or later given refinement. The extent to which just war thinking had become "disembedded" from the intellectual contexts that nurtured it – particularly those that were theological and religious – became especially clear to me when I received the text of a resolution regarding the current Iraq war that, in 2003, was put to the membership of the American Philosophical Association, the largest society (11,000+ members) of its kind for teachers and students of philosophy in North America. The first sentence of the rationale for the resolution, reads: "Both just war theory and international law say that states may resort to war only in self-defense." This statement simply presupposes, with no further exposition or argument, that the just war doctrine is a moral *lingua franca*, a common starting point whose terms and propositions are intelligible from whatever philosophical perspective the members of the American Philosophical Association may espouse, and a framework for moral judgment amenable to justification from a variety of moral perspectives and theories.

There is in my view, however, a double-edged consequence to this "disembedding" of the just war doctrine from the traditions of its originating thinkers. It certainly provides clarity to terms and structure of the theory, making them amenable to different lines of moral justification. That same clarity, however, may suggest to twenty-first century proponents of just war thinking that attention to historical contexts is relatively unimportant for understanding and applying the theory. Because the road across the centuries separating us from those earlier thinkers seems so easily traversed, we may come to think that we need not pay much attention to the intellectual and material dynamics that shape the larger framework of their – and our – thinking about matters of war and of peace and about the human persons and societies that are the agents of war and of peace. The apparent clarity of distinctions such as *ad bellum* and *in bello*, and the intuitive plausibility of governing criteria such as "legitimate authority," "right intent," "just cause," may make it easy to overlook the ways in which both our contemporary understandings of the overall shape of the just war doctrine and the understandings of our predecessors are deeply embedded in a larger nexus of concepts, theories, and accounts of the human condition – as well as in the human practices that shape those concepts, theories and accounts. The manner in which the just war doctrine has become, particularly in the later decades of the last century, a recognizable *lingua franca* for framing some of the important issues about war and peace, may, in fact, conceal an important two-sided interpretive dynamic. On the one hand, there is the extent to which even such a clearly stated doctrine brings in its wake a complicated heritage of theories of the role of the state and its governing institutions in securing conditions of peace for its citizens. On the other hand, there is the extent to which we on our part also bring to such a doctrine our own implicit and explicit contemporary accounts – often quite different from those within which just war thinking originated – of the same matters.

How and Why Context Matters: Material Conditions

What I thus would like to do in the next two sections is to offer what I hope will be an apt illustration of the importance of paying attention to the relationship between just war doctrine and the historical and cultural contexts in which its interpretation and use is framed. This illustration is drawn out of the discussions of the just war doctrine that, beginning in the 1980s and continuing periodically into the present, have been taking place within the US Catholic intellectual and church communities. These are discussions that have, from time to time during the past three decades, intersected with and occasionally stimulated larger public discussion on some of the issues in question, such as the moral status of various forms of nuclear deterrence policy, of the use of military force in humanitarian intervention, and of the rationales given in support of the legitimacy of the first and the second Gulf wars.

I consider these instructive because a major impetus that initially gave rise to them in the 1980s was precisely a recognition that the "material" conditions for the understanding and application of the just war doctrine had been significantly altered in a number of ways. The most notable of the material conditions that, at the time, was taken to require some rethinking of the just war doctrine – or at least of its application – was the advent of nuclear weapons and the consequences their development had for both strategic and tactical military planning – of which the most prominently discussed was the principle of deterrence (including that of "mutually assured destruction"). Attending to this dimension of the discussion is valuable for its potential bearing upon efforts to rethink just war doctrine in consequence of more recent alterations in some of the other "material conditions" in which war and other forms of armed conflict take place in today's world. Some of these altered material conditions are – like that of nuclear weapons – technological, while others – which I consider the ones that are more challenging with regard to rethinking of the just war doctrine – involve what appear as more fundamental shifts in the dynamics of the exercise of military power in relation to the political, social, economic and cultural structures that are undergoing transformation in an increasingly interdependent world.

It is with regard to this last point that I believe the ongoing discussion of just war within the Catholic community can be illustrative for discussion among the wider circle of just war theorists and their critics, because this Catholic discussion began – though by no means completed – an exploration of material conditions other than the technological that bear upon the reshaping of the just war doctrine. Although it did not figure as prominently as the question of deterrence did in the initial discussions – i.e., the ones that surrounded the drafting and issuing of the US bishops' 1983 document *The Challenge of Peace*[1] – I would argue that an equally important aspect of that discussion was the way in which it opened possibilities for rethinking the just war doctrine from the side of those larger accounts of the dynamics of the human condition in which just war thinking is embedded. In this case, the part of such larger accounts, which the ongoing Catholic discussion may usefully bring to our attention, concerns the material conditions provided by the structure and working of civil society. Among these are the role and responsibility of the sovereign power and its agents to promote the common good of civil society and its members and, conversely, the role of responsible judgment on the part of an informed citizenry with respect to the use of military force as means to achieve such good.

An Informed Citizenry and the Scope of Moral Judgment on Matters of "Just War"

Let me attempt to place this other aspect of the "material conditions" pertinent to the just war doctrine in a historical context. Just war principles bearing upon the decision to wage war, articulated to guide the exercise of prudential moral judgment of those who hold legitimate sovereign power in the state, took recognizable shape during eras in which monarchs and princes exercised such sovereign power. Just war principles bearing upon the conduct of warfare, articulated to guide those with responsibility for military planning and those in command of troops, took their shape in times during which combat and war making technology was mostly limited to fields of soldierly battle and their immediate environs. These contexts, however, no longer adequately delimit the structural and human circumstances in which those decisions are made for which just war principles are pertinent. The emergence of representative and democratic forms of government in the modern period, along with the concurrent rise of a more educated citizenry, the more recent development of more complex forms of regional and international economic and political interdependence – as much as the development of ever more powerful weaponry, increasingly rapid forms of military communication, transportation and logistics, and the growth of large standing armies as marks and instruments of sovereign power and national prestige – are matters that can and should have an impact upon the understanding, scope and the application of both main elements of the just war theory.

1 *The Challenge of Peace: God's Promise and our Response*. Specific citations from this document will be according to paragraph number

For purposes of providing a basis for further discussion beyond the brief treatment offered in this essay, I would like to call attention to one aspect of the altered "material condition" of the dynamics of civic and political life which, in a preliminary way, the US bishops' document, *The Challenge of Peace*, recognized for its potential importance in refining just war thinking. This refinement arises in response to changed circumstances in which modern representative and democratic forms of political order expanded the scope of citizen responsibility in the public realm. While reaffirming the *ad bellum* requirement that the *decision* to go to war is to be made by "competent authority," i.e., by the legitimate public authority of the state, the bishops also recognized that it was – at least under certain circumstances – within the scope of responsibility of individual citizens to make their own *judgments* concerning the fulfillment of the *ad bellum* criteria. A judgment that the criteria had not been fulfilled, moreover, could then be the basis for citizens legitimately to exercise a right of "selective conscientious objection" to military service in a war they deemed unjust. It seems to me – and this is a point certainly open to further discussion – that this empowering of moral judgment among the general citizenry with respect to the *ad bellum* criteria goes beyond the responsibility *to observe* the *in bello* criteria that the just war theory already recognized – again, within a set of larger conditions – as residing in the hands of individual combatants to refuse to follow orders involving conduct in violation of the *in bello* criteria. With regard to the latter, the bishops pointedly note that "to refuse to take such actions (that are unlawful or forbidden in warfare) is not an act of cowardice or treason but one of courage and patriotism."[2]

Extending the scope of a citizen's responsibility so that it now encompasses moral judgment with respect to the *ad bellum* criteria carries with it a number of important implications. One – which at least two points in the text of *The Challenge of Peace* suggest, but do not draw out in any detail[3] – is that this extension, in addition to authorizing selective conscientious objection for those faced with the prospect of induction into military service, also provides a basis upon which all citizens – not just those immediately liable to military conscription – may legitimately voice and organize public opposition against decisions engaging their nation in a war they deem unjust.

2 *The Challenge of Peace*, # 311.

3 One is # 324, which urges public officials to be "particularly attentive to the consciences of those who sincerely believe that they may not support warfare in general, a given war, or the exercise of a particular role within the armed forces"; the second, # 328, quotes, *Human Life in Our Day*, an earlier pastoral letter: "In our democratic system, the fundamental right of political dissent cannot be denied, nor is rational debate on public policy decisions of government in the light of moral and political principles to be discouraged."

I believe it important to make it clear that the issue that I think is most pertinent to further discussion is not that of the legitimacy of "selective conscientious objection." It is rather the significance of the range of "material conditions" that have an impact upon the shape of the just war doctrine, its interpretation and its use. While it seems to be the case that the most evident and significant of the material conditions affecting the just war doctrine since at least the middle of the 19th century have been technological advances that have exponentially increased the potential destructive range, power and speed of military operations, it seems – to me at least – that attention also needs to be paid to other material conditions, including the place for moral judgment by an informed citizenry as an element in the exercise of "competent authority," that shape the terms under which decisions to go to war are made.

Issues for Discussion

In consequence, I would like to offer the following set of considerations as a basis for further discussion:

1. An adequate historical understanding of the conceptual development of the just war doctrine requires that attention be paid to alterations in the most important of the full range of "material conditions" under which the doctrine originated and was later developed and refined.
2. Within the historical origin and development of the just war doctrine, one important material condition – at least in the period that stretches from Augustine through Aquinas to Suarez and Vitoria – was a set of religious and theological presuppositions bearing upon the (limited) role of the state and civil society in securing true human good for those whom they had the responsibility to protect. Included in these presuppositions – at least implicitly – was an understanding of the forms of justice and peace, which it was possible for civil authority to secure and to which the waging of war was ultimately ordered as its end. Articulation of these presuppositions about the securing of conditions for justice and peace to which the waging of war is ordered is likely to have important bearing upon discussions that have emerged about the inclusion of *post bellum* considerations as an integral element of the just war doctrine.
3. Many of the historically more recent developments of the just war doctrine – particularly those that place it within a larger conceptual frame work of international law (recurring, perhaps, to the earlier concept of *jus gentium*) – have not presupposed a religious or theological framework within which to place an articulation of the justice and peace to which the waging

of war might be ordered. There may, indeed, be implicit "anthropological" frameworks – i.e., accounts of what constitutes human good – that are embedded within these developments of international law and from which an account of justice and peace sought as the end of waging war can be articulated. Such frameworks, however, seem so far to have elicited relatively little attention as a significant part of the "material conditions" that need to be considered in shaping the just war doctrine. I consider it an important part of any project to revise the just war doctrine for the 21st century that it make a case that these "anthropological" frameworks (including those that place it within a religious/theological horizon) deserve significant attention as part of the "material conditions" for rethinking the just war doctrine. Such anthropological frameworks thus may also play a role in the articulation of the *post bellum* considerations noted in # 2, above.

4. Within the ambit of such anthropological frameworks, the dynamics and practices of political life – both internally and internationally – constitute another important set of "material conditions" for the shaping of the just war doctrine, its interpretation and its use. Recent developments of the just war doctrine – for example those articulated in papal documents since the time of John XXIII – have paid particular attention to the impact of a developing international order of interdependence as an important material condition upon the shaping of the just war doctrine.[4] One consequence of this has been – at least according to some interpreters – a trajectory toward further limitation of the circumstances under which any war might have legitimate claim to be "just."
5. In parallel to considerations of the material conditions upon the just war doctrine that are provided by a developing international order, attention also needs to be paid to the material conditions provided by the internal political and civil dynamics of particular societies. For instance, the organization, dynamics and practices of political life in a robustly functioning representative democracy, inasmuch as they constitute an important set of contemporary "material conditions" for the understanding of the just war doctrine and its application, require that we consider their impact upon how we might concretely and appropriately construe central notions in the just war doctrine such as "competent/legitimate authority" as well as the scope for the exercise of moral judgment by an informed citizenry with regard to the fulfillment of *ad bellum* conditions.[5]

---

4 Seminal documents for this development are the 1963 encyclical of John XXIII, *Pacem in Terris* and the 1967 encyclical of Paul VI, *Populorum Progressio*.

5 I wish to thank Michael Cumming for his helpful suggestions in preparing the final version of this essay.

References

Pope John XXIII (1963). *Pacem in Terris (Encyclical of Pope John XXIII on Establishing Universal Peace in Truth, Justice, Charity and Liberty).* [On-line]. Available: *http://www.vatican.va/holy_father/john_xxiii/encyclicals/documents/hf_j-xxiii_enc_11041963_pacem_en.html*

Pope Paul VI (1967). *Populorum Progressio (Encyclical of Pope Paul VI on the Progress of Peoples).* [On-line]. Available: *http://www.vatican.va/holy_father/paul_vi/encyclicals/documents/hf_p-vi_enc_26031967_populorum_en.html*

United States Catholic Conference (1983). *The Challenge of Peace: God's Promise and our Response. A Pastoral Letter on War and Peace*. National Conference of Catholic Bishops. Washington, DC.

United States Catholic Conference (1968). *Human Life in Our Day: A Collective Pastoral Letter of the American Hierarchy*. Washington, DC.

Asa Kasher

The Gaza Campaign („Operation cast lead") and the Just War theory[1]

Introduction

A properly functioning state should plan its actions carefully, execute them appropriately, and examine them scrupulously afterwards. A military operation is an important and complex act of state, and it is not exempt from proper planning, execution, and examination.

Whenever a state conducts a military operation outside of its borders, it engages in a political action. In a democratic state, the government must rigorously examine the political considerations and decisions that led to this action. A military operation also involves the deployment of armed forces and the cooperation of intelligence agencies. Each of these institutions is also expected to undertake a professional, methodical, and searching post-hoc inquiry into the considerations taken and decisions made in every professional locus of control that has had an effect on the operation – including those of operative planning, tactical performance, and intelligence. A responsible inquiry into these loci may then lead to a professional investigation of other loci of influence, such as those involved in capability building (i.e., developing military doctrines, practical training, etc.).

These political investigations and professional inquiries must pay special attention to every aspect of the operation that is related to moral and ethical values. Decisions, commands, and actions should be closely examined in order to determine whether they appropriately manifested the moral principles of the State of Israel, the ethics of the IDF and the General Security Service, and the laws to which Israel is subject.

There are two stages to such an analysis: First, one must determine the requirements that every military operation must fulfill in light of Israel's moral principles as a Jewish and democratic state;[2] the ethical codes of the IDF and the General

1 A brief version of the present paper was delivered in Vienna in a forum of the General Staff and the National Defense Academy. The present contribution is an abridged version of an article which was published in *Azure*, Summer 5769 / 2009, No. 37

2 For a deep presentation of the moral principles of the democratic state, see John Rawls, *A Theory of Justice*, revised ed. (Cambridge, Mass.: Belknap, 1999). The first edition was published in 1971, also by Belknap. The theory was reworked in John Rawls, *Justice as Fairness: A Restatement* (Cambridge, Mass.: Belknap, 2001). An elucida-

Security Service;[3] the laws that Israel must observe as a state in which the rule of law prevails; and the laws it must observe as a properly functioning state subject to both *jus gentium* (the "law of nations," i.e., international norms which apply to all states) and *jus inter gentes* (the "law between the peoples," i.e., treaties and agreements entered into by sovereign nations). Second, one must ascertain whether the decisions made, orders issued, and actions performed in the course of the operation fulfilled these moral, ethical, and legal requirements. In order to do this, one needs as reliable, full, thorough, and accurate an account of the relevant facts as possible. It is impossible to complete any moral, ethical, or legal evaluation of an operation before an investigation of its political background and an inquiry into the military's professional performance are completed. During and after Operation Cast Lead, many people, both in Israel and abroad, made statements about it as if this kind of examination had already been completed and its findings were at their disposal. Since it is reasonable to assume that not a single one of them had a reliable, full, thorough, and accurate account of the facts, their assertions can carry no moral, ethical, or legal significance at this stage.

For the time being, then, we should focus on the first stage of investigation mentioned above and restrict ourselves to examining the moral, ethical, and legal requirements to which decisionmakers and participants in military actions are bound. These requirements predate and are not dependant on the specific facts of Operation Cast Lead. However, though we are not in a position to provide a comprehensive answer to all the questions raised about what took place in the Gaza Strip during January 2009, the data collected so far permits us to conclude that a significant part of the criticism directed at Israel and the IDF during and after the operation was, to say the least, based on flimsy evidence.

---

tion of the principle of a "Jewish and Democratic State" based on Rawl's theory is found in Asa Kasher, "A Jewish and Democratic State: Philosophical Outline," *Iyyunei Mishpat* 19:3 (July 1995), pp. 729-739 [Hebrew]; this article was re-printed in Asa Kasher, *Spirit of a Man: Four Gates* (Tel Aviv: Am Oved and Yehoraz Association, 2000), pp. 13-27 [Hebrew]. The present paper can be well understood without usage of the Hebrew papers and documents mentioned in it. The latter are mentioned for the sake of Hebrew readers and completeness.

3 The most important of these are the "Spirit of the IDF" and the military code of ethics for fighting terrorism. See Asa Kasher, *Military Ethics* (Tel Aviv: Ministry of Defense, 1998), pp. 231-237 [Hebrew]. The "Spirit of the IDF" can be found at http://dover.idf.il/IDF/English/about/doctrine/ethics.htm. The values of the General Security Service can be found at www.shabak.gov.il/English/about/ Pages/valuseEn.aspx. For a specific discussion of the ethics of fighting terrorism, see Asa Kasher and Amos Yadlin, "Military Ethics of Fighting Terror: An Israeli Perspective," *Journal of Military Ethics* 4:1 (2005), pp. 3-32.

Just War Theory

The first factor one needs to consider when analyzing a military operation undertaken by a democratic state outside its own borders is the political decision to initiate the operation, and the circumstances under which this decision was taken. In order to evaluate such a decision properly, it is necessary to do so from two separate viewpoints: external and internal.

We will begin with the external viewpoint. From this point of view, we will morally evaluate the political decision to wage war or carry out a military operation based on considerations of international relations. The question we face at this point is the following: Does a state have a moral justification for taking military action against another state under the circumstances in question? We can pose similar questions about a military action against a non-state entity with a ruling body that, to a significant extent, effectively governs a specific territory (such as Hamas in the Gaza Strip); an organization that operates from within the territory of another state, and enjoys so much freedom of action that it effectively governs part of that state (such as Hezbollah in Southern Lebanon); and finally, an organization that operates from within the territory of a non-state entity (such as terrorist organizations operating in areas controlled by the Palestinian Authority). Even at this early stage of the discussion, it is already necessary to note not only the similarities between these situations, but also the differences between them each of which requires a separate moral discussion.

The basis for any such discussion is the moral conception known as *just war theory*.[4] This term does not designate a doctrine that has a single, authoritative form or interpretation. It is, rather, a family of concepts (e.g., "combatants" or "proportionality"), distinctions (e.g., between military and nonmilitary targets), and principles (e.g., that it is forbidden to harm enemy soldiers once they have surrendered). Moreover, scholars of just war theory often disagree about the meanings of its concepts, the considerations underlying its distinctions, and the specific consequences of its practical principles.[5] Over time, however, just war theory has developed an accepted set of eight principles, which form the basis of the stan-

4 For an important formulation of this theory, see Michael Walzer, *Just and Unjust Wars: A Moral Argument with Historical Illustrations*, fourth ed. (New York: Basic Books, 2006).

5 See, for example, Anthony Joseph Coats, *The Ethics of War* (Manchester: Manchester University, 1997); Paul Robinson, ed., *Just War in Comparative Perspective*, (Aldershot, England: Ashgate, 2003), parts I and IV.

dard moral discussion of war.[6] In addition, a framework of international law has emerged that constitutes the basis of customary legal discourse on the subject.[7]

Just war theory, as expressed in its moral principles and in international law, makes certain assumptions about the warring parties and the circumstances of their conflict. Usually, it is assumed that the warring parties are sovereign states and that, most of the time, circumstances permit a differentiation between combatants and non-combatants that is not too complicated. These assumptions, however, do not hold with regard to Israel's military confrontations with Palestinian terrorist organizations, Hezbollah in Lebanon, or the Hamas regime in Gaza. In order to apply the principles of just war theory to these engagements, it is necessary to widen its scope. For the purposes of the present discussion, however, we will abstain as much as possible from theoretical innovations, instead supporting our claims using the theory in its traditional form.

Just Cause, Last Resort

The first principle of just war theory in the circumstances under discussion is the *principle of just cause*. A state must have a compelling justification for taking military action against a state, entity, organization, or individuals outside its borders. From a moral standpoint, the only compelling justification for such action is self-defense. A state, therefore, can only justify military action if it can demonstrate that it acted on the basis of its *right to self-defense.*

No one can honestly dispute that, for years, Hamas, Islamic Jihad, and other terrorist organizations in Gaza have launched thousands of rockets at Israel's population. Therefore, we can present a responsible answer to the question: Was the decision to take military action against those terrorist organizations, at that particular time, justified on the basis of the right to self-defense? The answer is

---

6 The eight principles that are most commonly quoted in scholarly literature on just war theory are: legitimate authority, just cause, last resort, right intention, chances of success, macro-proportionality, micro-proportionality, and distinction. In this paper we will not consider the legitimate authority principle, because no one contests that the Israeli government was authorized to initiate Operation Cast Lead. We shall touch upon the other principles, also mentioning additional principles that are not normally considered part of just war theory, but nevertheless make important contributions to the moral analysis offered here. See, for example, Bruno Coppieters and Nick Fotion, eds., *Moral Constraints on War: Principles and Cases* (Lanham, Md.: Lexington, 2002).

7 See Yoram Dinstein, *War, Aggression, and Self-Defense*, third ed. (Cambridge: Cambridge, 2001); Yoram Dinstein, *The Conduct of Hostilities Under the Law of International Armed Conflict* (Cambridge: Cambridge, 2004).

self-evident: Firing rockets at Israel is an attack on the state and a constant endangerment of the life, health, security, and well-being of the citizens under attack.

Nevertheless, just war theory makes it clear that it is not enough for military action to be justified on the basis of self-defense. Though self-defense is a *necessary* condition for the justification of war, it is not a *sufficient* one. The moral considerations behind this assertion are clear: Military action poses a grave danger to human life, health, well-being, property, and liberty. If effective self-defense can be guaranteed by other means, this is clearly preferable to a course of action that involves destruction, suffering, and death. The use of military force is, therefore, justified only if all other alternatives have been exhausted. In just war theory, this is known as the *principle of last resort*.

Was the decision to launch Operation Cast Lead justified under the principle of last resort?

In order to answer this question in a responsible manner, it is necessary to understand the threats facing Israel's citizens, the possible alternatives to military action, the various attempts to pursue them, and the outcome of each attempt. Within the spectrum of threats, the rocket attacks initiated by terrorist groups – not only Qassam rockets but also longer-range missiles and more destructive weapons – must be considered. Alternatives to military action could include indirect negotiations for a ceasefire, international diplomatic pressure, and the imposition of a blockade. Israel, along with other international actors, pursued these options without success while Hamas and Islamic Jihad continued their rocket attacks on Israel's southern population.

The continued rocket attacks on Israel by the terrorist organizations in the Gaza Strip, as well as Israel's continued abstention from any large-scale military response in the face of this aggression, give rise to a presumption of justification regarding the state's decision to take military action as a last resort. Those who argue otherwise bear the burden of proof, and would need to demonstrate that: (a) there was a non-military alternative that Israel did not pursue; (b) had Israel pursued this alternative, its citizens would have been immediately and effectively protected from the threat of rocket attacks; and (c) this would have made military action unnecessary. To date, no alternatives that would have fulfilled these conditions have been proposed.

Some people claim that a peace agreement between Israel and the Palestinians would provide Israeli citizens with the best protection against rockets and missiles, suicide attacks, and other horrors of terrorism. It is true that a democratic state is required to seek peace agreements with neighboring states and peoples.[8] How-

8 See Asa Kasher, "The Democratic Obligation to Pursue Peace Arrangements," *Mehkarei Mishpat* [*Bar-Ilan Law Studies*] 14:2 (1998), pp. 343-361 [Hebrew]; this article was reprinted in Kasher, *Spirit of a Man*, pp. 145-171.

ever, the idea that it is possible to reach a political settlement with the Palestinians that would be upheld by Hamas, Islamic Jihad, and other terrorist organizations is quite doubtful. Even if we accepted the plausibility of such a claim, it is all but certain that rocket attacks on Israel would continue throughout the negotiations. In fact, they would likely *increase*. Leaving a state's citizens vulnerable to persistent threat is not morally justified by the mere fact of ongoing negotiations. Nor can the fact that negotiations are taking place justify avoiding the last-resort option after all alternative courses of action have failed. As long as a state's citizens are under attack, even during a negotiation process, that state has an obligation to provide them with adequate protection.

There are those who call on Israel to engage in direct negotiations with Hamas, in order to rid its citizens of the threats posed to them by rocket attacks and other kinds of terrorist activity. This argument warrants a similar response. From a moral standpoint, demanding that Israel engage in direct negotiations with a terrorist organization that does not recognize its right to exist cannot be justified. While indirect negotiations through some sort of mediation are a possibility, there is no basis for the supposition that this alone would be enough to achieve protection for Israel's citizens. Indeed, as we have already seen, rocket attacks continued while Israel was engaged in indirect negotiations with Hamas. Neither direct nor indirect negotiations can fulfill the three requirements mentioned above and cannot, therefore, be seen as effective alternatives to military action.

Right Intention

While a state entering into a war or embarking on a military operation must do so in self-defense and in the absence of other alternatives, these conditions alone do not suffice according to just war theory. A state may have other intentions – historical revenge, for example – which can alter the course of the war or its political aftermath, and which are not morally justified. Such motives can lead to excessive death and destruction, beyond what self-defense would require. The *principle of right intention* demands that a state not only wage war in a just cause, but that all of its intentions, on every level, be equally justifiable.

The aims of Operation Cast Lead included deterring Hamas and other terrorist organizations from launching rockets into Israel. Such deterrence is, in and of itself, morally desirable, because it can effectively prevent terrorist operations (or even war itself, as in the case of a state such as Syria). Nevertheless, measures taken in order to establish deterrence must meet certain moral requirements.[9]

---

9 See Lawrence Freedman, *Deterrence* (Cambridge: Polity, 2004).

The best method of achieving deterrence in a morally acceptable way is to achieve it as a side effect of some other action. Targeted killings of terrorists, for instance, not only offer immediate protection to a state's citizens. They also achieve deterrence, because the enemy becomes aware of the state's ability to detect threatening activities, identify the perpetrators and their whereabouts, and attack them. Deterrence is not the primary goal of targeted killings, however, but rather a welcome side effect. The intention to deter the enemy as a side effect of military activity constitutes a right intention.[10]

The difference between deterrence as a goal and deterrence as a side effect is essential. An operation whose goal is to thwart terrorist attacks should not be influenced by the likely possibility that it will also create deterrence. Theoretically, an operation can include the use of measures whose purpose is not to foil terrorist attacks but only to create deterrence. To the extent that injury or even death is caused as a result of these measures, they are morally unjustified. For example, killing someone who is essentially harmless in order to deter others from possibly posing a threat cannot be morally justified. A democratic state is required to protect human dignity as such. It cannot use human beings as mere tools to create deterrence. Human beings are not tools to be used.

Generally speaking, it is reasonable to ascribe to Operation Cast Lead the intention of achieving deterrence as a side effect of an act of self-defense. Likewise, descriptions of the operation as "disproportionate" in the Israeli and international media are problematic, because they appear to presume that deterrence was the main purpose of the operation, rather than a side effect of it. A description of the operation in terms of "powerful response" is more appropriate.[11]

Probable Success and Victory

One of the lesser-known principles of just war theory prohibits a state from embarking on a military campaign if it does not have a reasonable chance of winning it. War, by definition, involves the loss of human life, as well as suffering and destruction on a massive scale. The probability of success principle prohibits taking military action – which inevitably involves death, suffering, and destruction – if

10 For the Principle of Operational Deterrence, see Kasher and Yadlin, "Military Ethics of Fighting Terror," pp. 27-28.

11 The concept of "powerful response" has not appeared so far in the context of Operation Cast Lead, outside of specific references to plans for responding to future rocket attacks from Gaza. See Yarden Winter and Ido Elazar, "The Response of the Gaza Division to Rocket Attacks from Gaza: 'We Have Plans for a Powerful Response in the Gaza Strip,'" *Bamahane* 10 (March 6, 2009), p. 8 [Hebrew].

it is certain to fail. Therefore, it is impossible to justify a war that serves only a "symbolic" purpose.

This principle deserves our renewed attention in regard to military actions such as the Second Lebanon War and Operation Cast Lead, which were undertaken against guerrilla, terrorist, or terrorist guerrilla organizations. In conflicts of this kind, the definition of "victory" is different from that of "classic" wars between states and their armies. As we witnessed in the Second Lebanon War, a state's military action against a terrorist guerrilla organization can kill many of the group's combatants and destroy a significant part of its infrastructure without eliminating its ability to carry out terrorist activities and attack the state's citizens. During Operation Cast Lead, we realized that the same holds true in conflicts with urban terrorist organizations.

While the term "victory" may be emotionally satisfying, it is problematic from a professional military point of view. This is because it does not enable a clear distinction between the goals of a "classic" war and the goals of different kinds of wars or campaigns such as Operation Cast Lead. In these new contexts for the use of military force, it is best to replace the elusive term "victory" with the notion of "achieving specific goals by accomplishing the missions." This concept is easier to evaluate with precision and to use in professional employment of military forces.[12]

The primary goal of Operation Cast Lead was described as "improving the security situation" in the areas of the state under rocket attack. This is a proper objective, not only because it stems from the right to self-defense, but also because it is attainable. "Improving" the situation does not mean the elimination of all threats. An improvement can be attained by killing many terrorists, destroying much of their available weaponry, and causing heavy damage to their armaments infrastructure.

Proportionality (I)

"Proportionality" – a term raised many times in the context of Operation Cast Lead – actually refers to two different principles of just war theory: The *principle*

12 Pronouncements of this kind appeared in David Caspi, "Facing the Fences of Gaza in Amir's Armor Battalion," *Shiryon* 31 (January 2009) [Hebrew]. The Commander of the 75th Battalion in the 7th Armored Brigade, Lt. Col. Amir, was asked "what, in his opinion, would constitute a 'victory' over or a 'subjection' of Hamas." He answered that a "subjection" of Hamas would be achieved when "the enemy loses its ability and will to fight," while a "victory" would be achieved when "the battalion has successfully completed all of the missions it was charged with." The relationship between victory and subjection is complex and falls outside of the scope of our discussion.

of macro-proportionality, which applies to the overall decision to take military action, and the *principle of micro-proportionality*, which applies to specific military actions. I will now turn to the first principle, and address micro-proportionality later in our discussion.

In order to clarify the issue, we must examine some of the common-place accusations of disproportionality made regarding the Second Lebanon War and Operation Cast Lead. The most common charge raised by critics of these campaigns concerns the number of casualties. They argue that, since very few people are killed by rocket attacks on Israel's population, while many people are killed by the Israeli response, this response is disproportionate and, therefore, both morally unjustified and contrary to international law. This claim is both invalid and groundless. It is invalid because the number of Israeli casualties is not a reliable measure of the threat posed by enemy rockets. Let us recall the Grad rocket that hit an Ashkelon classroom on February 28, 2009, which happened to be a Saturday morning. Had the missile hit the school on a day when classes were in session, dozens of school children would have been killed. The good fortune of these children does not diminish the threat posed by the attack itself. A responsible comparison between Hamas attacks and Israeli action during Operation Cast Lead would not distinguish between "hits" and "close calls." It would take into account the thousands of rockets that have been fired into Israeli towns and cities, and would reach the conclusion that the Palestinian threat to Israeli citizens is greater than the Israeli threat to residents of the Gaza Strip who reside in the vicinity of the terrorists.

Furthermore, no principle of proportionality entails a demand for numerical equivalence. A moral evaluation of proportionality in military action should focus on the question of whether the positive results of the operation on one front outweigh the negative results on another. Macro-proportionality requires that this condition be met. The positive results of the operation should be measured in terms of the protection it has provided to the state and its citizens at the conclusion of the military campaign and its aftermath. The negative results should be measured in terms of the death, suffering, and destruction inflicted on the other side. Once again, this is not a numerical comparison, but rather an assessment of existing threats and the measures that must be taken in order to avert them.

Let us examine, for example, the circumstances of the outbreak of the Second Lebanon War. During the first stage of the war, Hezbollah fighters killed eight Israeli soldiers and kidnapped two who died of their wounds. At this point, Israel was faced with two threats: First, that Hezbollah would carry out another operation in which they might succeed in killing or kidnapping more Israeli soldiers. Second, were Israel to take military action in order to avert the first threat, Hezbollah might respond by barraging the north of Israel with Grad rockets. In order to protect itself from both of these dangers, Israel needed to launch a strike against the sources of the second threat, both in South Lebanon and in certain neighborhoods of South Beirut. Whether or not Israel could have minimized the damage inflicted on a specific site in South Beirut or any other area without diminishing the security of its citizens is a legitimate question. But most of the accusations of disproportionality during the Second Lebanon War were more generalized and thus invalid and unsubstantiated.

Similar charges were made about Operation Cast Lead. For example, the Spanish author Javier Marías told the Israeli newspaper Haaretz that Israel "drew a gun in response to a slap in the face."[13] This is a colourful expression, but it is also an invalid and unsubstantiated claim. The cause of Operation Cast Lead was not a "slap in the face," but a very long series of thousands of slaps. After how many slaps and attempted slaps and threats of slaps is it time to draw a gun? I assume that the author did not mean to say that Israel should have turned the other cheek, so it seems reasonable to assume that he thinks Israel should have responded to a slap with a slap, or a punch at most. He thus missed the major point: Israel's response was not undertaken simply for the sake of responding, but to obtain genuine protection for itself and its citizens. One is permitted to ask: How can we know that an Israeli slap would prevent the next Hamas slap? It seems reasonable to assume that even a punch would not have prevented future slaps or lowered the threat they posed – which is not the case with "drawing a gun."

Moreover, accusations that Israel "drew a gun in response to a slap in the face" entirely miss another essential point: In this context, proportionality is not assessed by simply comparing the Israeli military response to a specific enemy operation ("the slap"). Instead, it involves a comprehensive assessment of Operation Cast Lead in light of the ongoing actions that the enemy has committed for many years and will continue to commit for the foreseeable future ("endless repeated slaps"). Such an assessment should take into account the enemy's desire and ability to inflict continuous harm on Israeli citizens. After all, Israel did not draw a "gun" in response to one "slap in the face," but in response to constant slapping, attempts to slap, and threats to inflict stronger and more powerful slaps.

Accusations of disproportionality in war often refer to the "use of excessive force." To justify these claims, one would need to offer alternatives in which the use of force (a) would not be excessive; (b) would be effective, i.e., provide the required protection from specific threats;[14] and (c) would be available when the circumstances require it. Considering these conditions, it is not surprising that we

13 Roi Bet Levi, "King Javier the First," *Haaretz*, March 1, 2009 [Hebrew].

14 In order to assess effectiveness, one would need to rely on intelligence information that is, by its very nature, not widely accessible.

have yet to encounter any defensible criticisms of the use of overwhelming force. Indeed, they are quite difficult to make from the comfort of one's armchair.

Internal Justification

So far, we have discussed the principles of just war theory from the external viewpoint, which focuses on the interaction between the state and outside bodies and forces. Now we will turn to the internal viewpoint, which is concerned with the relations between the state, its institutions and basic arrangements, and the citizenry.

The most important aspect of the relationship between a state and its citizens is the *obligation of self-defense*. This is one of the highest duties of a properly functioning democratic state. It requires the state to protect its citizens – indeed, anyone under its effective control – from any danger to their life, health, security, and well-being resulting from acts of violence, both in the short and long term. As a democratic state, it must fulfill this obligation with proper respect for the human dignity of all people.

The distinction between the external viewpoint and the internal viewpoint becomes apparent when one considers the difference between a state's *right* of self-defense, which relates to what is beyond its confines, and a state's *obligation* of self-defense, which relates to what is within its confines.

A state must protect its citizens from acts of violence, whether from a foreign state or from a terrorist or guerrilla organization. This obligation is binding whether its citizens are threatened by an external source or an internal source; whether the cause is criminal activity or political subversion. The obligation of self-defense is based is a simple rationale: A democratic state is characterized by a system of fair arrangements of civic life. In order to uphold this system, the state must preserve and defend the conditions that enable it to exist. The most important of these conditions, without which the citizen cannot enjoy the arrangements of democracy, is the very fact that the citizen is alive. A democratic state is therefore under an obligation to defend its citizens' lives. (The same principle guides the state's obligation to defend its citizens' health, security, and well-being. For the purposes of this discussion, however, we will not deal with these considerations, and will restrict our analysis to life-threatening dangers.)[15]

15 For a discussion of the "infrastructure principle," the "principle of preserving human life," and the "principle of protecting the lives of soldiers," see Kasher, "The Democratic Obligation," pp. 155-164.

A state's obligation of self-defense grants each of its citizens the right to ask it the following question: "What have you done to protect me from a given violent threat that endangers my life?" (The *citizen's question.*) Every citizen has the right to receive a satisfactory response to this question, a response that will refer him to the institutions, arrangements, policies, and actions that are charged with protecting him against the threat he has in mind.

The Soldier's Question

A state is never exempt from its responsibility to give a satisfactory answer to the citizen's question. This also holds true when the citizen is serving in the military, for the simple reason that soldiers are citizens. The state owes them a satisfactory answer just as much as it owes one to every other citizen. At times, the *soldier's question* will be more challenging and the state's answer will be more complex. Usually, a properly functioning state does not intentionally design a situation that will endanger the lives of its citizens. When a citizen is put in harm's way, the state ought to defend that citizen in an effective manner. However, a citizen in military service may find himself in an extremely dangerous situation because the state has knowingly sent him to risk his life on its behalf. The soldier's question will therefore be twofold: "First, what justification do you have for sending me into a life-threatening situation? Second, once I am in this situation, what are you doing to protect me from the danger I am in?"

We will not give a full account here of the state's response to the soldier's question. Such a response would have to justify conscription and reserve military service, insofar as they are rooted in the fair arrangements befitting a democratic state under present conditions. Instead, we will only mention one key component of Israel's reply to the first part of the soldier's question: "We have no choice," or, in other words, "It is *necessary* to do so under the circumstances in which we find ourselves." Because of the threats facing Israel and its citizens, the state cannot fulfill its obligation of self-defense without imposing conscription and reserve military service. When a threat to its security is not imminent, the state is required to develop its military capabilities. When the threat is imminent, it must exercise its military power. While preparing for conflict, the state places severe restrictions on the liberties of its uniformed citizens. When using military force, the state may send them into battle and thus endanger their lives. "Being forced by circumstances," the state must impose conscription and reserve military service in order to fulfill its obligation of self defense.

Proportionality (II)

Just war theory distinguishes between *jus ad bellum* and *jus in bello*, that is, between the moral justification for war – which we have already discussed – and the moral justification for actions taken during a war. The question of moral conduct in war, upon which we will now focus, must be evaluated according to the proper relationship between the state and what is outside of it (the external viewpoint), as well as between the state and its citizenry, including its soldiers (the internal viewpoint).

The distinction between *jus ad bellum* and *jus in bello* is manifest in the difference between the principle of macro-proportionality explained above, and the principle of micro-proportionality to which we now turn. Just war theory requires that all actions conform to the principle of proportionality, not only with regard to the decision to wage war or a military operation, but also in regards to specific military actions that endanger harmless enemy non-combatants. Similar to macro-proportionality, the principle of micro-proportionality concerns the question of whether or not the positive consequences of actions on one front morally justify the negative consequences on another.

It is easy to answer this question affirmatively when military action, in terms of both its goals and the means of achieving them, is unavoidable. In other words, the action is of *military necessity* in the strict sense of the word.[16] The aim of such an action is to fulfill the state's absolute duty to defend its citizens, given the dangers they face. The means employed to meet this requirement must be those which can most successfully fulfill the obligation to protect the citizens of the state as well as the human dignity of all people. When security conditions make it a necessity, military action accompanied by a genuine effort to minimize harm to enemy non-combatants can be justified under the micro-proportionality principle, because it positive consequences outweigh the negative ones.

Many military actions, however, do not fall strictly within the scope of military necessity. Often, the means required to carry out various actions are not, in a sense, unavoidable. Sometimes a military commander can choose between achieving the mission's objectives through a difficult, slow, and problematic process, and doing so in a simple, fast, and easy way. Let us assume that these two options do not diverge in terms of the degree to which they endanger the lives of the soldiers involved, but only in terms of the length of time and the magnitude of the effort required to achieve the objective. The micro-proportionality principle demands

---

16 For a discussion of different, problematic uses of the concept of "military necessity," see Paul Christopher, *The Ethics of War and Peace: An Introduction to Legal and Moral Issues*, second ed. (Upper Saddle River, N.J.: Prentice-Hall, 1999). The strict use of the term in this paper is free from the problems inherent in other uses.

that the positive consequences of employing the faster and less demanding option justify its negative consequences, namely, inflicting death, suffering, and destruction on enemy non-combatants. There is, obviously, no ready answer to the question of which option is preferable in some cases, because we usually possess only a partial picture of the facts, and have to take into account multiple factors and conditions. Take, for example, situations in which soldiers are required to carry out a specific mission and, afterward, must continue immediately to another urgent and difficult mission. If both missions are militarily necessary, then it is preferable for them to carry out the first mission in what they consider the easiest way, despite the fact that it may be more harmful to enemy non-combatants. On the other hand, if the soldiers do not expect the first assignment to be followed by another urgent and complex mission, then it is better for them to take the more difficult course of action, thus causing less harm to enemy non-combatants.

In order to know whether the micro-proportionality principle was upheld during Operation Cast Lead, it is necessary to be familiar in a full and detailed way with the specific actions taken during the operation. One cannot judge the operation in a serious, professional, and responsible manner without having adequate knowledge of the actions in question, and one should therefore resist the political and emotional temptation to do so.

Principles of Distinction

Just war theory also demands that combatants respect the *principle of distinction.* This is a key principle in moral discussions of military actions, and it ought to be properly understood. A crude and superficial presentation of the principle of distinction often creates a slippery slope that leads to conclusions that cannot stand up to moral scrutiny. Therefore, we will exercise special caution in presenting and explaining it. Though the main elements of the principle of distinction were formulated with the classic concept of war in mind, they will be described here so as to be applicable to the "newer" context of fighting terrorism.[17]

The principle of distinction presents the combatant with three different standards of conduct to guide him in any military action: (a) a standard he should follow when facing a group comprising enemy combatants and no one else; (b) a standard he should follow when facing a group of enemy non-combatants who are not participating in the fighting and are not in proximity to enemy combatants;

17 For novelties in these contexts, see Paul Gilbert, *New Terror, New Wars* (Washington D.C.: Georgetown, 2003), ch. 1, pp. 1-23.

(c) a standard he should follow when facing a mixed group of combatants and non-combatants.

It is important to understand that we are not drawing a distinction between different kinds of people, but rather between different standards of conduct to be applied in different situations. The first standard of conduct permits soldiers to attack enemy combatants freely without considering the immediacy of the danger they pose – with the exception of wounded persons, prisoners of war, medical teams, and clergy.[18] The second standard of conduct prohibits attacking enemy civilians who are not involved in hostilities and are not in proximity to enemy combatants. This restriction is *absolute*. Under certain conditions that we shall elucidate at length, the third standard of conduct permits attacking enemy combatants even if this endangers non-combatants in their vicinity.

The moral rationale behind the principle of distinction, which institutes multiple standards for military action, is self-evident: Military conduct that complies with the principle of distinction greatly reduces the horrors of war. Nevertheless, the question must be posed: Does this principle possess a deeper moral justification? Furthermore, is the framework of standards that it establishes of the highest moral standing, or is it simply superior to circumstances in the past, in which armies freely and equally harmed combatants and non-combatants? This question reveals a fundamental dispute that need not be resolved here in order to evaluate Operation Cast Lead.[19] Even those – and I am among them – who hold that the standards of conduct delineated by the principle of distinction do not offer an ideal moral solution to the problem will nevertheless respect them, and seek to replace them with arrangements that are better both in theory and practice.

Operation Cast Lead mostly took place under conditions that required the application of the third standard of conduct, as well as considerations of micro-proportionality. The above-mentioned third standard enables us to answer the difficult question of what should be done when dealing with a group of people that includes both terrorists who pose a threat to the safety of Israelis and enemy non-combatants who do not threaten anyone. In a situation such as this, we face a dilemma: If the terrorists remain unharmed, they will continue to threaten Israelis. Attacking these terrorists, however, is likely to injure or even kill their non-combatant neighbors. Either way, people who should not be harmed and whom the circumstances of combat do not justify harming will be hurt.

---

18 There are also other reservations, including some that relate to the means and tactics of combat, but they are not relevant to the present discussion.

19 An attempt to justify the principle of distinction on moral grounds, based on the special status of combatants, is found in Michael Walzer's *Just and Unjust Wars*. For the opposite claim, see Asa Kasher, "The Principle of Distinction," *Journal of Military Ethics* 6:2 (2007), pp. 152-167.

The first way to attempt to resolve a dilemma is by altering the situation so that there will be no need to choose between alternatives, all of which involve undesirable consequences. In the situation we just described, it would be necessary to separate the people who pose a threat from those who do not. Efforts to do this may include scattering leaflets notifying people about the impending attacks, contacting specific places by phone in order to issue a warning, using non-lethal weapons, etc.[20] If enemy combatants and non-combatants are successfully separated, there is no need to use the third standard of conduct, since the first standard will be employed against the terrorists and the second will protect their neighbors from being injured. The trouble is that, despite all efforts, such a separation is not always possible, frequently because warnings would alert the terrorists to a coming attack and thus make it more difficult to defend people from them. What, then, should be done when the dilemma cannot be eliminated, and soldiers are faced with a heterogeneous group of hostile terrorists and harmless non-combatants?

The third standard of conduct allows combatants in such situations to make a double effort: They should try to ensure that they strike the terrorists with high probability, and they should try to minimize harm to harmless civilians. Whenever these two demands are incompatible, the first is preferable to the second, but never overrides it entirely.

Discussions of just war theory relate the above-mentioned third standard of conduct to the *double effect principle*. According to this principle, when we are seeking a goal that is morally justified in and of itself, then it is also morally justified to achieve it even if this may lead to undesirable consequences – on the condition that the undesirable consequences are unavoidable and unintentional, and that an effort was made to minimize their negative effects. Micro-proportionality is also a required condition.

Thus, civilian casualties – though an undesirable, painful, and troubling reality – are an acceptable outcome of a military action if they cannot be avoided. During Operation Cast Lead, it was claimed that it is prohibited to attack one hundred terrorists if one child might be harmed along with them. This claim is both morally indefensible and utterly irresponsible. No one wants to harm a child, but refraining from attacking one hundred terrorists because of the child they hold

---

20 In the past, the so-called "neighbor procedure" was also used, in which the IDF would ask a neighbor or relative of the terrorist to approach the terrorist – without any risk to their own person – in order to ask the terrorist to hand himself over to the IDF. This procedure has since been banned by a Supreme Court ruling. In my opinion, an improved, more appropriate, and secure version of this procedure should have been authorized in order to offer more protection to the lives of all parties involved in the conflict. See HCJ 3799/02, *Adalah – The Legal Center for Arab Minority Rights in Israel et al. v. Commander of the Central Command, IDF et al.*, ruling from June 23, 2005.

means allowing them to continue attacking Israeli civilians – including children. Is it justified to allow a child – or an adult, for that matter – to be harmed in Israel in order to avoid harming a child in Gaza?

Israeli historian Zeev Sternhell raised an equally weak argument when he severely criticized Operation Cast Lead and warned that "the Israelis... by wreaking havoc on a civilian population... remove themselves from the family of civilized Western nations."[21] In response to this claim, we should first recall that Israel was fighting Hamas, which is considered a terrorist organization by many among the "family of civilized Western nations." Destruction, suffering, and casualties are, unfortunately, inevitable during urban warfare against such groups. The moral question, then, is whether the destruction, suffering, and casualties were justified in light of the continuing attacks by Hamas and other organizations against Israel and its citizens. The mere fact that destruction, suffering, and casualties were inflicted in the Gaza Strip in no way answers this question.

Secondly, if we wish to evaluate Israel's place among the "family of civilized Western nations" in the context of "wreaking havoc in a civilian population," we might learn a great deal from comparing Operation Cast Lead to Operation Phantom Fury, which the United States Marines launched in the city of Fallujah, west of Baghdad, during November and December 2004. According to a report published by the United States Army Combat Studies Institute, many of Fallujah's 350,000 residents fled the city before the operation, leaving an estimated number of 3,000 insurgents behind. During the operation, about 6,000 Iraqis and 1,200-2,000 insurgents were killed. Of the city's 50,000 buildings, some 10,000 were destroyed, including 60 mosques, each of which was used to store substantial quantities of armaments and munitions. Over half of the city's buildings were substantially damaged.[22]

The Gaza Strip is home to a population five times larger than the number of people that lived in Fallujah prior to Operation Phantom Fury, and about twenty times larger than the population that remained in the Iraqi city after the mass flight. The number of terrorists in Gaza was more than five times the number of insurgents in Fallujah. We do not yet know with certainty how many people were killed in Gaza, but even the terrorist organizations and their supporters do not claim that it was five times the number killed in Fallujah – which would be 30,000 dead. We

21 Zeev Sternhell, "Kadima to Democracy's Rescue," Haaretz, March 6, 2009 [Hebrew]. This is not the place to discuss the meaning of the phrase "the family of civilized Western nations," but it would also be inappropriate to entirely ignore its condescending and even racist tone. For a discussion of non-Western theories of just war, see Robinson, *Just War in Comparative Perspective.*

22 Kendall D. Gott, *Breaking the Mold: Tanks in the Cities* (Fort Leavenworth, Kan.: Combat Studies Institute Press, 2006), pp. 91-106.

also do not know with certainty how many buildings were destroyed in Gaza, but no one claims that the number is anywhere close to 50,000. Thus, a simple calculation shows that the United States, a senior member of the "family of civilized Western nations," left a trail of destruction in Fallujah that was at least twenty-five times greater than anything Israel inflicted on Gaza during Operation Cast Lead.[23]

It is important to note, however, that such a comparison cannot serve as a valid basis for reaching moral conclusions. American actions in Fallujah and Israeli actions in Gaza may both be morally justified; and, possibly, neither of them is morally justified. It is also possible that Operation Cast Lead was justified, but Operation Phantom Fury was not, or vice versa. I mention the American operation only to demonstrate that there is no basis to the claim that Israeli conduct of anti-terrorist urban warfare is worse than that of other enlightened states.

We have emphasized the fact that harmless civilian casualties may be an unavoidable consequence of military action in defense of a state's citizens against terrorists. How does one defend the claim that a particular outcome is "unavoidable"? The first step is to review the various courses of action that were available to the military forces. The second step is to assess the effectiveness of each of these options in terms of the probability of successfully striking the terrorists. Third, it is necessary to evaluate the potential consequences of each course of action that are both possible and undesirable. These considerations should point us toward the preferable approach. For example, attacking a building in which dangerous terrorists and harmless civilians are present is likely to cause harm to non-combatants even if the preferred method of action is used. Harming these civilians would be considered unavoidable only if all other alternative courses of action are less desirable, whether because they present an even graver danger to the terrorist's neighbors or because they are ineffective against the terrorists themselves.

The task of examining possible courses of action in terms of their potential consequences, both desirable and undesirable, with respect to the above-mentioned moral principles (the principle of distinction and the double effect principle), should be reserved for professionals. In regard to one point, however, an

23 The number of casualties and the scale of destruction caused by Operation Cast Lead as reported in Israeli and foreign media are not reliable, in my opinion. Lorenzo Cremonesi, a correspondent for Italy's *Corriere della Sera* and the only foreign reporter to enter Gaza during the operation, presented evidence that called into question any numerical assessment coming from Gaza. This skepticism is also based on past experience: One should recall the use of deceitful propaganda by Palestinian terrorist organizations and their supporters when they accused Israel of a "massacre" in Jenin. The number of casualties they cited was grossly inflated in comparison with the number confirmed by an objective investigation. See "Palestinian Civilians in Gaza to 'Corriere della Sera': Hamas Wants Israelis to Shoot at Our Homes," *Haaretz*, January 29, 2009 [Hebrew].

additional moral consideration is involved: Imagine that a military force is dealing with a situation in which dangerous terrorists and their harmless civilian neighbors are inside the same building. Let us assume that the military force arrayed against the terrorists has already invested considerable efforts in attempting to separate enemy combatants from non-combatants by issuing warnings in writing, by phone calls, by loudspeaker announcements, by using non-lethal weapons, etc. Nevertheless, a mixed group of terrorists and their harmless neighbors still remain in the building. We will also assume that the preferred course of action, from a professional military point of view, is to fire at the building from the ground or from the air, which is likely to harm the terrorists and some of their neighbors. At this juncture, it would be apparently reasonable to offer an alternative course of action: sending soldiers into the building in order to separate the terrorists from their neighbors. If this is successful, then the soldiers can retreat from the building and the attack will only target the terrorists. If, however, the soldiers' attempt to separate the terrorists from their neighbors is unsuccessful, then the military force will have no choice but to fire at the building despite the possibility of harming non-combatants. It is self-evident that such a course of action presents a greater risk to the lives of the soldiers who are to be sent into the building than a course of action that does not send them in.

There are those who claim that this is a proper risk if it will reduce civilian casualties. In my opinion, however, there is no moral justification for favoring the lives of a terrorist's neighbors over the lives of soldiers. In what follows, I shall raise three points to support my position. First, we must recall the moral challenge of the soldier's question: "I am a citizen of the state. I enter combat in uniform because it is my duty to participate in defending other citizens of the state from a danger they face. It is for lack of any other choice that I am put into dangerous situations in order to attack terrorists and thereby defend my fellow citizens. These dangerous situations inevitably threaten my life. To send me into a building in order to reduce the chances of injuring enemy non-combatants will significantly increase the threat to my own life, and not for the purpose of carrying out the mission of eliminating terrorists, but rather for the purpose of protecting their neighbors. What justification is there for increasing the threat to my life?" I am not aware of a compelling response to the soldier's question and the moral challenge it presents. I am familiar with several attempts to formulate such a response, but they are unsuccessful, unpersuasive, and do not justify endangering soldiers' lives in this manner.[24]

---

24 I will mention one example: In the past, but also more recently, some commentators claimed that the value of human dignity requires that the human dignity of neighbors of terrorists should be protected. At the same time, these commentators described soldiers as state instruments. Such an approach deprives soldiers of their human dignity and is

Second, the suggested solution we just described is a good example of the slippery slope down which a careless presentation of the principle of distinction can lead. If this principle is used in order to draw a distinction between different types of people, it will inevitably lead to preferring some people over others in a sweeping and unjustified manner. An accurate presentation of the principle of distinction would avoid this slippery slope. Although we have two different military standards of conduct – the permissibility of attacking combatants during war and the impermissibility of attacking non-combatants – we cannot deduce from these a third standard of conduct that requires a state to prefer protecting the lives of enemy civilians over the lives of its own soldiers.

Third, and perhaps most important, we must consider the *special duties principle*. While morality demands that the human dignity of *all* people be protected, the provisions regulating this protection are not necessarily universal. Canada's obligation of self-defense, for example, is Canada's obligation toward Canadian citizens just as Israel's obligation of self-defense is Israel's obligation toward Israeli citizens. Canada does not have a duty to defend Israelis who are not in Canada, and Israel does not have a duty to defend citizens of Canada who are not in Israel. Israel is expected to offer assistance when a natural disaster hits another state because of its obligation to protect the human dignity of all people, but it is obviously required to invest greater efforts when such a disaster hits within its own borders than when it hits elsewhere. Israel's special obligations toward its citizens far exceed its duties toward all human beings as such.

Israel, like every other democratic state, is bound by a hierarchy of duties toward different populations. On the first, and highest, tier of this hierarchy, we find Israeli citizens. Just below them, on the second tier, are residents of the state who are not citizens, such as permanent residents, foreign workers, visiting tourists, etc. These are all the people who are found within Israel's international borders (the "Green Line"). On the third tier are the residents of the territories over which Israel has had effective control since the Six Day War and who are not Israeli citizens. On the fourth tier, which lies far below the preceding tier, are residents of territories that Israel does not effectively control and are not Israeli citizens. There is a decisive difference separating the first, second, and third tiers from the fourth. Israel is accountable for actions within its borders and in territories that it

therefore invalid. Soldiers are not instruments, but rather uniformed citizens, and the obligation to respect human dignity as such applies to them as well. Examples of this problematic attitude can be found in Colm McKeogh, "Civilian Immunity in War: From Augustine to Vattel," in Igor Primoratz, ed., *Civilian Immunity in War* (Oxford: Oxford, 2007), pp. 62-83. For a contemporary example, see p. 80.

effectively controls; it is not accountable for actions that take place in territories in which it has no effective control, such as Gaza, Greece, or Canada.[25]

Where do Israeli soldiers stand in this hierarchy? Being Israeli citizens, they belong to the first tier of the state's hierarchy of duties. When they are not serving in the IDF, they are full citizens and the State of Israel is duty-bound to them, just as it is to all of its citizens. However, when they are in military uniform and engaged in military activity, special obligations and restrictions imposed on them lower their place in the hierarchy to somewhere between the third and fourth tiers. The state has a duty to protect its non combatant citizens and the rest of the people under its responsibility with its uniformed, combatant citizens; consequently, the lives of its soldiers are often jeopardized. It is important to emphasize that the state must come up with a compelling justification for endangering the lives of its soldiers. In principle, it should do so only under the necessity of self-defense.

Therefore, in the dilemma at hand, the state should favor the lives of its own soldiers over the lives of the neighbors of a terrorist when it is operating in a territory that it does not effectively control, because in such territories it does not bear the responsibility for properly separating between dangerous individuals and harmless ones. Once it has exhausted its efforts to separate terrorists from non-combatants, not only is the state no longer obligated to endanger the lives of its own soldiers in order to attempt to further such a separation, it is forbidden from doing so. (Some micro-proportionality considerations have to be mentioned in the full description of the decision we recommend, but since they do not change our practical consequences we will not presently describe them.)

The position presented here is also formulated in the IDF's ethical code of conduct.[26] Though the values outlined in the code, known as the "Spirit of the IDF,"

---

25 The implications of the concept of "effective control" are manifold. See Sari Bashi and Kenneth Mann, "Control and Responsibility: The Legal Status of the Gaza Strip after the Disengagement," *Hamishpat* 25 (May 2008), pp. 13-24 [Hebrew]. Despite the claims presented in that article, it is clear that Israel does not have effective control over the Gaza Strip in a way that renders it responsible for separating between terrorists and their non-combatant neighbors.

26 One should recall that the "Spirit of the IDF" is a summarized version of the IDF's first ethical code, "The Spirit of the IDF: Values and Basic Norms." The original document included a two-part definition for each value, one concise and one longer. *Purity of arms* was defined there as follows (notice that the term "soldier" in the IDF context applies not just to the army, but also to the air force and navy): "The IDF soldier will use his weapons and force to the extent necessary for subduing the enemy and will exercise restraint in order to avoid causing undue harm to human life, person, dignity, and property. The IDF soldiers' purity of arms constitutes the restrained use of weapons and force while carrying out the missions, only to the extent required for accomplishing them, without undue harm to the human life, person, dignity, and property of soldiers or civilians, especially helpless dependants, during war and everyday security operations,

are presented in an abstract and abridged form, it is nevertheless possible to derive clear conclusions from them with regard to our discussion.

First, the definition of the value of *purity of arms* in the IDF code states: "A soldier will not use his weapon and force to harm non-combatants or prisoners of war." In line with the double effect principle mentioned above, when a soldier uses his weapon to attack a terrorist and unavoidably harms non-combatants at the same time, he is not using his weapon "to harm non-combatants." He is acting the way he does in order to harm other, dangerous individuals, whom it is his duty to attack under the circumstances, in order to protect his fellow citizens from them.

Secondly, the value of purity of arms includes the demand that IDF soldiers "will do all in their power to avoid causing harm to the lives, bodies, dignity, and property [of non-combatants]." In order to properly understand the phrase "do all in their power," let us examine the following example: Terrorists have taken over a residential building in Gaza, owned by a harmless non-combatant. Only terrorists are present in the building. They can be attacked in two ways: One possibility is to bomb or shell the building and topple it with its present inhabitants – who are solely terrorists – still inside. The other possibility is to send soldiers in to kill the terrorists without demolishing the building. The first option will, of course, "cause harm to… [the] property" of the non-combatants who live there, as opposed to the second possibility, which will not cause the same degree of destruction, but will gravely endanger the lives of the soldiers involved. The professional and moral solution to this problem is self-evident: It is not acceptable to risk soldiers' lives in order to avoid damaging the property of non-combatants.

Such a decision coincides with the demand that soldiers "do all in their power," because a soldier's courses of action are not determined by the physical options available to him, but rather by professional ethical considerations. The limits placed on a soldier's conduct are dictated by the values and principles that he must uphold. Among these is the *value of human life*, which asserts that, "During combat, [a soldier] will endanger himself and his comrades only to the extent required to carry out their mission." When a soldier enters a building, his mission is to attack the terrorists inside. He thus puts himself at risk. If he is a commander, he

---

in armistice and peacetime." *Human life* was defined there as follows: "The IDF soldier will act strongly to save human life, out of recognition of the supreme value of human life, and will endanger himself or others only to the extent required to carry out the mission. The sanctity of life as upheld by IDF soldiers will be expressed in all of their actions, in judicious and careful planning, considerate and cautious training, and proper performance in accordance with the mission, the professionally appropriate extent of danger and extent of caution, and will manifest a continuous effort to lower the number of casualties to the extent absolutely necessary for accomplishing the mission." See Kasher, *Military Ethics*, p. 232.

endangers his soldiers "to the extent required to carry out the mission." A soldier will not endanger himself or other soldiers in order to avoid damaging the property of enemy civilians.

The value of human life delineates the limits of a soldier's conduct not only in regard to damaging the property of non-combatants, but also to harming their "lives, bodies, [and] dignity." We have already seen that soldiers are required to endanger themselves only "to the extent required to carry out the mission." Soldiers are *not* required to endanger their own lives in order to reduce the risk of harming a terrorist's neighbors. They "will do all in their power to avoid causing harm" to non-combatants, but without risking their own lives and the lives of their comrades.[27]

Over the course of Operation Cast Lead, questions arose regarding situations in which soldiers' lives were endangered due to causes other than enemy action. For instance, four Israeli soldiers fell in the course of the operation in "friendly fire" incidents. In another case, IDF soldiers were wounded by an Israeli mortar shell that landed near their position.[28] In a third incident, one Israeli force fired at another, though the episode ended without casualties.[29] There is no need to repeat here that such incidents are unwarranted. At the conclusion of the operation, when asked how to avoid "friendly fire" incidents, Colonel Ilan Malka, commander of the Giv'ati Brigade, noted the importance of the value of *professionalism*, and immediately added an insight into the appropriate attitude toward soldiers' lives. He said, "We have to explain [to the commanders] that even if they lose a bunch of terrorists, it is no big deal. Don't shoot if you are not sure you know where your troops are." In this context, Malka pointed out a clear failure in the preparations for the operation. "We did not discuss this enough when going over the procedures of the fighting.... It did not come up as much as the other topics came up. We went in without being sufficiently prepared on this issue."[30] "Friendly fire" is not unavoidable, and some of the confusion that happens during combat is indeed un-

---

27 On one occasion I was told that my position implies a demand for "zero risk to the lives of soldiers." This claim is groundless. When a soldier operates in a battlefield against terrorists, he is endangered by enemy sniper fire, small arms, grenades, mortar shells, rockets of different types, roadside bombs, loader bombs, booby-trapped buildings, assault and kidnapping threats, etc. Everyone agrees that the soldier is expected to remain in such a dangerous situation during a military operation.

28 Daniel Al-Peleg, "My Travels in Hamastan," *Bamahane* 6-7 (February 6, 2009), pp. 29-34 [Hebrew].

29 Nir Costi, "Champagne," *Bamahane* 6-7 (February 4, 2009), pp. 103-105 [Hebrew].

30 Col. (Res.) Saul Nagar, "I Insist that the Ninth Battalion Join Me in This Battle," *Shiryon* 31 (January 2009), p. 45 [Hebrew]. The article covers a joint operation of the Giv'ati Brigade and the 9th Battalion of the 401 Armored Brigade; Col. Amir Eshel, "The Military's Force is in Ground Maneuvering," *Shiryon* 31 (January 2009), p. 47. Eshel, Head

necessary. This confusion can be ameliorated, and Israeli casualties from "friendly fire" can thereby be reduced. When it comes to military action, the value of professionalism requires showing proper respect for the value of *human life*.

Another danger that faced Israeli soldiers during Operation Cast Lead was the possibility of being kidnapped by the enemy and being used later as a bargaining chip against their own country. IDF soldiers are supposed to be trained in precautionary steps to prevent abduction and to properly respond to kidnapping attempts. The Hannibal Procedure Rules of Engagement, drafted prior to 2000 while the IDF was still present in Lebanon, instruct a soldier how to act professionally in order to preempt an attempted abduction. Unfortunately, this order was misinterpreted by both commanders and the media, who believed that it sanctions the killing of a soldier, either by his own hand or at the hands of his comrades, in order to prevent him from being taken alive by a terrorist organization. This is, of course, unacceptable both ethically and morally. The purpose of the Hannibal Procedure is first and foremost to retrieve the abducted soldier and return him safely to his home and army unit. It is inconceivable that a military order would require IDF soldiers to kill one of their comrades or require an Israeli soldier to commit suicide when abducted.[31]

Conclusion

All military actions carried out during Operation Cast Lead should undergo a professional, thorough, and detailed investigation, just like any other non-routine and complicated professional operation. Moral, ethical, and legal evaluations of specific actions can only be undertaken through the methodical and systematic framework of a professional inquiry. Only on the basis of these inquiries will it be possible to arrive at a general conclusion regarding Operation Cast Lead. Allegations of war crimes against Israel or its soldiers, which started at the time of the operation and continue today – long before these inquiries have been completed and the lessons drawn from them made public – are neither objective nor serious. The ease with which they are made serves as sufficient proof that they draw their inspiration from the deceitful propaganda of Israel's enemies.

I shall conclude my essay with two comments. In his *Nichomachean Ethics*, Aristotle wrote, "it is for a noble end that the brave man endures and acts as courage directs," which, according to a Jewish commentator, refers to "those who en-

---

of the Armored Corps Doctrine Development Center, defined the friendly fire incidents as the IDF forces' most crucial failure.

31 See the basic rules 24-25 in "The Spirit of the IDF: Values and Basic Norms," in Kasher, *Military Ethics*, p. 236.

danger their lives in war to save the community."[32] I am deeply impressed with the courage displayed by each and every one of the soldiers who participated in Operation Cast Lead and their commanders. They acted and suffered "for a noble end," and endangered their lives "in war to save the community." I would like most of all to commemorate the four officers and six soldiers who died in combat, along with the three civilian casualties, and to pay my respects to their families. At the same time, I am deeply grieved on behalf of each and every one of the harmless Palestinians who were not involved in terrorism, but nonetheless died during the operation due to the malicious designs of Hamas.

References

Al-Peleg, D. (2009, February 6). My Travels in Hamastan. *Bamahane 6-7*, 29-34.

Amir, E. (2009, January) The Military's Force is in Ground Maneuvering. *Shiryon 31*, 47.

Aristotle (1867). *Book of Ethics, Stanav* (p. 26a). Lvov: I.O Salat.

Aristotle (1963) *The Works of Aristotle, Ethica Nicomachea, book III* (p. 1115b). Oxford: Oxford University.

Bashi, S. & Mann, K. (2008, May) Control and Responsibility: The Legal Status of the Gaza Strip after the Disengagement. *Hamishpat 25*, 13-24.

Caspi, D. (2009, January). Facing the Fences of Gaza in Amir's Armor Battalion. *Shiryon 31*.

Christopher, P. (1999). *The Ethics of War and Peace: An Introduction to Legal and Moral Issues*. Upper Saddle River, N.J.: Prentice-Hall.

Coats, A. J. (2003). *The Ethics of War*. Manchester: Manchester University.

Coppieters, B. & Fotion, N. (Eds.) (2002). *Moral Constraints on War: Principles and Cases*. Lanham, Md.: Lexington.

Costi, N. (2009, February 4). Champagne. *Bamahane 6-7*, 103-105.

Cremonesi, L. (2009, January 29). Palestinian Civilians in Gaza to 'Corriere della Sera': Hamas Wants Israelis to Shoot at Our Homes. *Haaretz*.

Dinstein, Y. (2004a). *The Conduct of Hostilities Under the Law of International Armed Conflict*. Cambridge: Cambridge.

Dinstein, Y. (2004b). *War, Aggression and Self-Defense*. Cambridge: Cambridge.

Freedman, L. (2004). *Deterrence*. Cambridge: Polity.

32 Aristotle, *The Works of Aristotle*, trans. W.D. Ross, vol. 9 (Oxford: Oxford University, 1963), *Ethica Nicomachea*, book III, ch. 7, p. 1115b; Aristotle, *Book of Ethics*, translated and elucidated by Itzik Stanav (Lvov: I.O. Salat, 1867), p. 26a [Hebrew].

Gilbert, P. (2003). *New Terror, New Wars* (pp. 1-23). Washington D.C.: Georgetown.

Gott, K. D. (2006). *Breaking the Mold: Tanks in the Cities* (pp. 91-106). Fort Leavenworth, Kan.: Combat Studies Institute Press.

John, R. (1999). *A Theory of Justice*. Cambridge, Mass.: Belknap.

John, R. (2001). *Justice as Fairness: A Restatement*. Cambridge, Mass.: Belknap.

Kasher, A. (1995, July). A Jewish and Democratic State: Philosophical Outline. *Lyyunei Mishpat, 19 (3)*, 729-739.

Kasher, A. (1998a). *Military Ethics* (pp. 231-237). Tel Aviv: Ministry of Defense. Available: http://dover.idf.il/IDF/English/about/doctrine/ethics.htm

Kasher, A. (1998b). The Democratic Obligation to Pursue Peace Arrangements. *Mehkarei Mishpat, 14 (2)*, 343-361.

Kasher, A. (1998c). *The Spirit of the IDF: Values and Basic Norms. Military Ethics* (p. 236). Tel Aviv: Ministry of Defense.

Kasher, A. (2000). *Spirit of a Man: Four Gates* (pp. 13-27). Tel Aviv: Am Oved and Yehoraz Association.

Kasher, A. (2007). The Principle of Distinction. *Journal of Military Ethics, 6 (2)*, 152-167.

Kasher, A. & Yadlin, A. (2005). Military Ethics of Fighting Terror: An Israeli Perspective. *Journal of Military Ethics, 4 (1)*, 3-32.

Levi, R. B. (2009, March 1). King Javier the First. *Haaretz*.

McKeogh, C. (2007). Civilian Immunity in War: From Augustine to Vattel. In I. Primoratz, (Ed.), *Civilian Immunity in War* (pp. 62-83). Oxford: Oxford.

Nagar, S. (January 2009)I Insist that the Ninth Battalion Join Me in This Battle. *Shiryon 31*, 45.

Robinson, P. (Ed.) (2003). *Just War in Comparative Perspective. Parts I and IV*. Aldershot, England: Ashgate.

Sternhell, Z. (2009, March 6). Kadima to Democracy's Rescue. *Haaretz*.

Walzer, M. (2006). *Just and Unjust Wars: A Moral Argument with Historical Illustrations*. New York: Basic Books.

Winter, Y. & Elazar, I. (2009, March 6). The Response of the Gaza Division to Rocket Attacks from Gaza: 'We Have Plans for a Powerful Response in the Gaza Strip'. *Bamahane 10*, 8.

Juha Mäkinen

Creating a Unified Framework for Future-Oriented Education of Military Ethics

Introduction

The study aims at producing a normative framework for educating officers in military ethics, applicable at military universities and colleges. The study goes beyond mere case studies of best practices in teaching ethics and the excellent analyses of how ethics *is taught* at national military colleges and universities (e.g. see Perry 2005a; Lagacé-Roy 2007), although it has been influenced also by these inspiring studies. The intent is not to explicitly reflect on the Finnish experiences in teaching ethics, either. Instead, the approach is *prescriptive* and *normative* in the sense of paying attention to *ought to*- and *should*-type questions in the field of military ethics by creating a one-dimensional ethical framework to be collaboratively and globally elaborated in the near future.

The main premises of the paper

The first main premise of the present paper is that the military ethical studies written by the military ethicists of the IAMP, can and need to be utilized when the big puzzle, or a coherent framework, for educating officers in military ethics will be put together. It will be scrutinized how clear a direction for ethical education is given in our military institutions by the current military ethical studies. No matter what kind of shape the answer for this question may take, it seems, nevertheless, to be reasonable to assume that the analyzed studies will have, at least in "a hidden" and "unlinked" form, many crucial patterns and pieces to be used for the "guiding" puzzle and framework. Consequently, in order to fill the "clarity" requirement, the created framework consists of only one dimension or axis, at least for the start.

The second main premise of this paper is that in the globalizing world it is extremely important that military ethics can be developed on the basis of the ideas of "human security", bioethics, and global ethics (Toiskallio 2007, 11), but also with close resonation with other forms of applied ethics and social sciences in general. We often seem to assume that we are "doomed" to the vicious circles of unmaintainable relations to the world or to our realities in general (cf. Värri 2007, 37), and this inevitability is often assumed to be caused by some faceless global economic forces, namely by our highly unethical business organizations, which are totally out of our control. Here it is assumed that not even our business organizations, and not necessarily our war-fighting or crisis management ones, are doomed to be unethical. But as we all actually are well aware of, also business can become unethical when it forgets *its point and purpose* (Solomon 1992: 1993, 20; cf. e.g. Ghoshal 2005; cf. Toiskallio 2004, 114 about the "good" for military science) and when *the development of moral character* (e.g. Kohlberg 1984; Colby & Kohlberg 1987) is overridden by selfish and short sighted views. However, in practical terms, in the midst of these kinds of global challenges the business ethics seems to have kept its relevancy also for the managers of multinational enterprises and corporations operating worldwide. Interestingly, the relevancy has been kept even despite of the overt "prescriptive" and "normative" stance balancing the more "practically", "empirically", and "descriptively" oriented stances of business ethics and managerial practices.

In this paper the interdisciplinary turn, namely the turn towards e.g. business ethics (i.e. contractualism; cf. e.g. Heugens, Kaptein, van Oosterhout 2006) is taken due to the fact that our current global and planetary challenges will be solved, if they can be solved at all, through collaborative inter-cultural and transdisciplinary[1] double hermeneutic processes[2] (Giddens 1984, 1990, 2001; cf. also Habermas 1984).

A more theoretically oriented way of justifying the turn towards "inter- and transdisciplinary military ethical studies" is an obvious need to make sense of the various "Kuhnian paradigms" (1962: 1970), "schools of thought" and "Lakatosian research programmes" (Lakatos & Musgrave eds. 1970; Lakatos 1978) of ethics, applied ethics, and military ethics, while *not focusing solely on virtue ethics* (e.g. Aristotle; MacIntyre 1985; cf. also e.g. Solomon 1992:1993; cf. articles in Toiskallio 2007; Toiskallio 2003, 99) *or alternatively on Kantian deontological considerations only* (e.g. Royl 2002; Royl forthcoming a; b; articles in Toiskallio 2007; Toiskallio 2003, 101-102).

A utilitarian, or more precisely put, consequentialistic stance[3] is currently emerging among us in the IAMP, and also contractualism will get some more

1 Here transdisciplinarity means research involving researchers from two or more disciplines, and also practitioners. Interdisciplinarity, in turn, means a research involving two or more scientific disciplines, being a form of transdisciplinarity (cf. Gibbons, Limoges, Nowotny, Schwartzman, Scott & Trow 1994; Nowotny, Scott, Gibbons (2001).

2 Meaning interaction between the meaningful social life of the practitioners (i.e. the military practitioners; the soldiers) and the social scientific worldview.

3 See Attfield (2003, 190) for a brief definition of consequentialism. Cf. e.g. Talerud 2007; Verweij 2007; Nissinen 2007, 107; Lagacé-Roy 2007, 110, 112; Mutanen 2007, 144-146; Heinonen 2007, 205.

proponents in the midst of the current planetary challenges and security threats (cf. e.g. Attfield 2003; cf. Toiskallio 2007; Talerud in Toiskallio 2007; Henk in Toiskallio 2007). For the soldiers and officers alike this all means in practical terms that they can not any longer escape their responsibilities for their actions and activities by claiming to be "just obeying orders as soldiers", they are also citizens, voters and other kinds of influential decision-makers in our globalizing democracies.

The created ethical framework allows our officers to learn (i.e. learn deeply) while making ethical considerations along the dimension of micro (culturally specific local features) and macro aspects of the global reality. Here it is fully agreed with Etienne Wenger (1998, 141; cf. also Mäkinen 2006) that knowing in practice involves an interaction between the local and the global. Consequently, in the unified framework the main axis or dimension is the one between local culturally specific and relativistic features *balanced* by universal and global concerns[4]. Although for a start the created framework will be a simple one-dimensional one, it does not allow making the traditional conclusion of neglecting our main paradoxes (cf. Värri and Verweij in Toiskallio 2007[5]; cf. e.g. Stadler 2003, 9) and dilemmas (cf. Lagacé-Roy in Toiskallio 2007; Toiskallio 2007; Annen 2007; Raviv 2004), or turning a blind eye to our situational complexities when continuing to reduce the ethical "education" on the level of espoused codes of conduct to be memorized and putting the scope on the "universal right answer" (cf. Lagacé-Roy 2007, 113; cf. the levels of moral development by Kohlberg (1984; Colby & Kohlberg 1987).

The third main premise of the present paper is that officers studying military ethics tend to have (at least in Finland, but potentially elsewhere as well) the needed intellectual curiosity to progressively inquire into the conclusions with authentic cases[6] balanced by normative concerns on a global scale. Finally, at least almost needless to say, the teachers of ethics at our university also have the needed courage to let the students inquire into the "practical depths" of our current paradoxes and challenging dilemmas along the micro-macro dimension.

4 I.e. hypernorms. E.g. human dignity is often emphasized also in the codes of conduct of the Armed/Defence Forces and it is a first candidate to be an example of a hypernorm (e.g. Henk 2007, 230; Perry 2005a, 156; Lagacé-Roy 2007, 114; Talerud 2007, 70; Cohen, Libel 2007, 179).

5 Namely, "how is it possible to maintain one's ethical consistency and be an ethical subject by justifying killing and respecting human dignity at the same time?"

6 Examples of such authentic cases are e.g. the hidden "Eichmannian stance of ourselves" (MacIntyre 1998, 200); Hugh Thompson in My Lai; Lieutenant Ehren Watada and major Florian (Royl, forthcoming a) in the case of "Operation Iraqi Freedom" operations and cases gained from peace-keeping and crisis-management missions (Lagacé-Roy 2007; Canada 2006).

Current interpretations of the military ethicists of the IAMP

This section follows a path and cognitive trail[7] through the field of military ethicists of the IAMP while identifying the main axes and dimensions revealed by the analyzed studies. The analyzed military ethical books are *Military Pedagogy – an International Survey* (2002; edited by Heinz Florian), *Cultures, Values and Future Soldiers* (2002; edited by Jarmo Toiskallio, Wolfgang Royl, Reijo E. Heinonen and Pekka Halonen), *Civil-Military Aspects of Military Ethics* (2003; edited by Edwin R. Micewski), *Identity, Ethics, and Soldiership* (2004; edited by Jarmo Toiskallio), *Military Ethics in Professional Military Education – Revisited* (2005; edited by Edwin R. Micewski and Hubert Annen), and *Ethical Education in the Military: What, How and Why in the 21st Century?* (2007; edited by Jarmo Toiskallio).

As it is immediately clear, the intent is not to give a full coverage of the military ethical studies done by the members of the IAMP but due to the basic nature of the current project (= a progressive inquiry done in an open-ended manner), instead of being some kind of a curse, it will be a blessing if the following conclusions are elaborated after adding more ethical articles of the IAMP members to the ones to be analyzed. Even more remarkably, the further inquiries will not be a curse even in such a possible, although not a probable case, where the basis of the created framework will be shaken by the extended analyses and reconsiderations.

Due to the fact that even the selected six books contain a vast amount of information, knowledge and even wisdom, it is helpful to use a holistic model as "glasses" while aiming at identifying some hidden patterns, axes, and dimensions in the books. Figure 1 shows the holistic model in question, namely the hermeneutical military-pedagogical planning model (Toiskallio 2003, 119), when put into use in ethical education in the military.

7 Adrian Cussins' theory of cognitive trails (Cussins 1992; 1993; 2003) is an analytical tool to make sense when we are "walking through our reality and world while making sense of it".

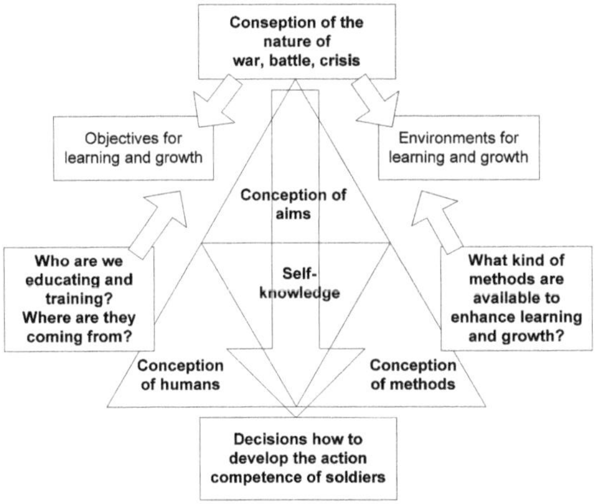

Figure 1. The hermeneutical military-pedagogical planning model

In the present analysis some main features of the holistic model are emphasized: the conception of the nature of war, battle, crisis and security threats has been rapidly changing, extending "the area of interests" to be considered by every soldier. The main competencies, values and expectations of the students to be educated and trained are changing, giving additional competencies to be utilized in the educational processes of the Armed/Defence Forces. The available educational methods and means for military educators are changing and should be changing further in the military educational institutions due to the advancement in social scientific research, the beliefs and expectations of the student population, and the technological means to support learning by e.g. web-based tools. The identity of the officers of Armed/Defence Forces is changing, needing to be reconsidered in nationally oriented Armed/Defence Forces.

According to Desirée Verweij (2007, 59) when speaking globally, military ethics seems to be operative at three levels of our reality: micro, meso and macro. The expansion of awareness has been a rapid one, as many of us have recognized, and the consequences have been enormous on all the different levels of military ethics. Micewski (2003, 41) explains macro effects as follows:

However, with the corresponding alteration of the global geopolitical design of security, western military organizations have changed from mere instruments of national defence mechanisms of international crisis prevention and conflict resolution.

When putting more flesh on the "one-dimensional" bones of the created framework we *should be* aware not just of the continuously gained experiences from various kinds of missions, but of *potential missions of the future.* In practical terms this means that although ethical education should consist of traditional Just War Theories (e.g. Walzer 2000; Perry 2005a, 159; cf. Pangle & Ahrensdorf 1999) and current missions (e.g. Pitt & Ritter 2002; Winer 2004[8]), it has to go beyond these (e.g. Kasher & Yadlin 2005; cf. Perry 2005b), including "potential missions" in e.g. Iran (for an brief introduction see e.g. Macgregor 2003; Ansari 2006), North-Korea (Carpenter & Bandow 2004; McCormack 2004), and also "potential approaches" to the future security threats (e.g. deterrence versus preventive and preemptive approaches; see e.g. Dershowitz 2006).

Despite the macro level chaotic turbulences, the indispensability of the micro level actors plays an ever-increasing role in the contemporary dramas of our lives. The importance of the individual level has everything to do with the fact that the individual must first of all develop moral and ethical competence within him or herself, before being able to develop it on the other levels (Verweij 2007, 59; cf. Toiskallio, 2007; 2004). In other words the moral and ethical competence *potentially* emerges on the micro level of our existence and *critical reflection is a crucial ingredient for the emergence to happen* (cf. Verweij 2007, 59; Fleck 2005, 71-72; Lagacé-Roy 2007, 111).

It is clear that the moral competence discussed above cannot be achieved with regulation alone. It seems that we need to educate every soldier, despite of their rank and position in our organizations acknowledging that everyone is a leader; and hopefully a wise one (cf. Peter Foot (2003) in Toiskallio 2004, 117). For military ethics this line of thinking means that although military practitioners tend to require codes of conduct to be followed on field missions, we as military ethicists have to be fully aware of the dangers of reducing military ethics to a level of publishing codes of conduct only, to be distributed via an electronic means for our soldiers to be memorized in their "tabula rasas" (cf. Stadler 2003, 3-4 about the "tabula rasa" point of view) or "downloaded into their minds".

The identities of our soldiers are in transition (Royl 2002; cf. Royl 2004; Toiskallio 2003, 103), shifting from mere national identities towards Euro-Military identities or even to global identities in times of the rise of global ethics and global concerns of human security (Henk 2005; 2007; Kalevi Mäkinen 2005; Toiskallio 2007; Talerud 2007). One fundamental question for military ethicists

8 It is really worth asking and figuring out whether we in the West (US) project our domestic problems on to the world stage and what kind of a role is played by the US (Western) military-industrial complex having vested interests with our (US; European) political elite?

and practitioners alike is how to enable such an identity transition through for example the European Bologna processes?[9]

As emphasized above, the question of shared, but also personal values and moral interests comes to the fore. This stance can be captured in the form of pivotal questions: what do I really stand for, what do I find important, and why? (cf. Verweij 2007, 59; cf. Lagacé-Roy 2007, 111) The time has come to ask how justifiable it is in our current global situation to claim that the officers' moral bond of obligation *only* exists between himself, the nation, the other soldiers as peers and other kinds of national ideals.

As we currently often tend to argue, one of the main justifications for our collective crisis management efforts is to safeguard our shared values, an unpleasant, but a highly essential question may pop up on your personal HUD screen: what are those shared values not just in the west but also in the east and the south? As Nissinen (2007, 106) highlights "there can not be true commitment without shared values". I would like to emphasize that here Nissinen is not talking about some kinds of "shared espoused values" but more or less of "shared actual values" truly deserving to be guaranteed. Instead of seeing different value patterns in cultures as a source of problems (Ercil & Varoglu 2002, 102) it is here claimed that they are a fact (cf. e.g. Hofstedes 2005[10]), but in what kinds of issues do the values differ for example between the Americans and Europeans (cf. e.g. Royl 2004, 71)? It seems to be definitely possible to identify some value conflicts, but how to handle these kinds of conflicts – turn a blind eye to them and forget the "value conflicts" or pay attention to them? Instead of seeing potential value conflicts as "bad", it is justifiable to see them as valuable resources to be utilized in learning processes – especially in expansive learning processes (Engeström 1987; Mäkinen 2006).

An intriguing question deserves our full attention. If we agree with e.g. Micewski, who claims that "if one agrees with the fact that any military officer – in an appropriate understanding of the concept – should belong to the intellectual "elite" of a country's society, there still is a long way to go (Micewski 2005, 20)." Is there really? How culturally and nationally varying this claimed aspect actually is or is not depends both on national military-political decisions shaping our military educational institutions and on the teaching practices in the military educational institutions. Partly, it depends on us! In the midst of the complexities of our even broader area of interests it seems to be justifiable to ask whether we are recruiting for our Armed/Defence Forces personnel who have the needed potential to become members of our intellectual "elite" and who have crucial capacities for

9 See http://ec.europa.eu/education/policies/educ/bologna/bologna_en.html; Mäkinen 2005.

10 According to Hofstedes (2005, 4) *human nature* is what human beings have *in common* despite of their cultures and personalities.

“quality education” (Toiskallio 2007, 6) (i.e. university level, highly qualified education that is both theoretically and practically oriented).

According to Guenther Fleck (2005, 70), two different modes of teaching ethics can be distinguished: Teaching ethics via *rational argumentation* (third person perspective) and teaching ethics via *sensitizing* (first person perspective; cf. also Lagacé-Roy 2007, 117). The first mode represents the classical way of teaching by focusing on well-elaborated philosophical concepts. The second mode represents a newer way of teaching by focusing on the student’s *emotional* experience. In the first mode students learn something about ethics; in the second they *become familiar with ethics*. Conclusively, it can be said that military ethicists have traditionally favoured teaching ethics via rational and well-elaborated philosophical and social scientific argumentation.

We must now turn our attention towards the “second mode” of teaching ethics that has emerged in the field of military ethicists of the IAMP and also within the military practitioners. But before we continue further, one crucial point deserves to be highlighted. Instead of some sort of “revolution”, we are talking about a normal cultural evolution where *both* modes of teaching ethics are in a *balanced* and *synthesized resonation within each other*. Finally the needed turn is not only among us as military ethicists of the IAMP but in line with Toulminian thinking (cf. 1990, 186) the “modern” focus on the written, the universal, the general, and the timeless need to be broadened to include once again *and also* (=let us not forget theories and ethical books) the *oral* (=spelling out of reasons for your own actions and behaviour), the *particular* (while avoiding the trap of overriding cultural relativity by keeping our hypernorms in our minds), and the *timely* (=what is really our current situation right now; right here on this planet?). It is claimed here that the major turn contains a minor turn towards practical trans- and interdisciplinary military ethical studies where applied ethicists (i.e. business ethicists) supply us with equipment to construct a bridge across the divide between military ethics and ethics; and to the side of the soldiers. In what follows, the construction of the bridges will be explained.

Introduction to the Integrative Social Contracts Theory of Thomas Donaldson and Thomas Dunfee

During the past ten years the Integrative Social Contracts Theory (ISCT; Donaldson, Dunfee 1994; 1995; 1999; 2000; 2001; 2002; Dunfee 2006[11]) has emerged as a promising integrative approach in the field of decision-oriented contractualist business ethics. Instead of a broad overview of the details of the ISCT, its basic elements are introduced below, followed by some recommendable reviews of the ISCT (see e.g. Fort 2000; Boatright 2000; Rowan 2001) and also some of its critics (e.g. Hartman, Shaw, Stevenson 2003) to be utilized by progressively inquiring military ethicists.

Since the times of Aristotle (384-322 BC), the relationship and tension between *the pole of local and particular* (what "is" in terms of G.E. Moore (1873-1958); the descriptive and empirical stance) and *the pole of universal* (what "ought to be" in Moorean terms; the normative and prescriptive stance) have been in tension for us (MacIntyre 1985, 148; MacIntyre 1998, 244). The problem has often been called "the naturalistic fallacy", meaning that one cannot logically derive an *ought* from an *is* (MacIntyre 1998, 244). Being deeply aware of the fundamental tension between these poles of interests, Donaldson and Dunfee (1994) have identified how the field of business ethics, too is divided by two *complementary* approaches (Donaldson, Dunfee 1994, 253; Donaldson, Dunfee 1999, 9). ISCT is an explicit attempt to integrate the two rival camps of methodology meaningfully and connect ethical considerations with the realities and complexities of practices (cf. Donaldson, Dunfee 1999, 166).

According to Donaldson and Dunfee, at the most basic level, if all that ISCT achieves is to influence managers to consider the interests (i.e. norms, values etc.) of relevant communities and the existence of universal moral principles, it will have had a positive impact (Dunfee 2006, 304). Hence, ISCT *balances* an overtly culturally relativistic stance by so called hypernorms but it does not neglect local norms and values either (cf. Donaldson, Dunfee 1999, 23). Metaphorically speaking and analogically to swinging a golf club, where both hands are needed, in the ISCT approach two main elements are needed: namely authentic locally shared norms and values but also hypernorms (Dunfee 2006, 306; cf. Hartman, Shaw, Stevenson 2003).

The first major question posed by the ISCT approach is "what the authentic norms of the community in question are"? In order for a norm to be an authentic

11 During the research project (see Mäkinen 2006) the ISCT was recognized as an interesting attempt to make sense of this crucial micro-macro dimension and the ISI Web of Science was a valuable search-tool for identifying the research referring to the article (1994) and to the book (1999) of Thomas Donaldson and Thomas Dunfee.

one, both attitudinal and behavioural consensus of the norm are required within the community in question (Dunfee 2006, 316). This means that in this approach norms are not imposed to micro-level actors but instead the actors themselves specify what the authentic norms of the community really are and the duty of "outsiders" is to find them out. The shared dialogic process (Dunfee 2006, 321; Calton 2006) of the actors involved acts as a means to overcome bounded moral rationality while evoking appropriate moral intensity.

As already mentioned, the ISCT approach includes *the option to constrain* the local authentic norms. Hence, it acknowledges a limit to that local legitimacy (Donaldson, Dunfee 2001, 436). One rule of thumb of the ISCT is that transactions solely within a single community, which *do not* have significant *adverse effects* on the other humans or communities, should be governed by the host (=local) community's norms and values (Donaldson, Dunfee 1999, 184[12]). This means that the ISCT approach does not aim at constraining local communities in every case and issue, but it pays attention to the *consequences* of actions and to those being affected by the actions. In other words, the ISCT approach reminds us of the long-term consequences and does not let us rest with the shortsighted view. Additionally, and in philosophical terms, while the ISCT also focuses on the consequences of the actions, it at least implicitly, if not even explicitly, shows its "consequentialistic" and pluralistic stance by denying "the necessity" to be either a utilitarian, or a virtue ethics theorist, or a deontologist, etc. (cf. (Donaldson, Dunfee 1999, 23), but not at all putting ethical theories into the bin. Among the business ethicists the question over hypernorms has remained a frequently debated issue, but has it been addressed by the military ethicists and soldiers? One candidate for a hypernorm was identified above (cf. footnote 35) – human dignity[13]. Let us see how applicable it is in our missions.

A Practical Application of the Military Ethical Framework

First of all, the ISCT approach enables us as military ethicists and military practitioners to ask **"fundamental questions"** instead of easy secondary ones (Măkinen 2005) in the field of military ethics. The first "fundamental question" is the one

12 There are six rules of thumb in total (Donaldson & Dunfee 1999, 184-190; Donaldson & Dunfee 2000, 491). Cf. e.g. Moses if he had come down from the mountain with 128000 commandments and the consequent reduction of credibility and impact (Dunfee 2006, 307).

13 The reader is reminded of the "categorical imperative" of Immanuel Kant (1724-1804) as "Act so that you treat humanity, whether in your own person or in that of another, always as an end and never as a means only." (cf. Royl, forthcoming b).

about the authentic local norms of the community in question. The second "fundamental question" is the one about hypernorms identifiable in the codes of conduct, ROEs, military policies and in many kinds of strategies. Both these fundamental questions deserve to be progressively inquired by the teachers and students of military ethics.

For a start, a potential hypernorm needs to be identified: let it be human dignity. The progressive inquirers have to identify authentic cases in the field from various kinds of missions and contexts while checking whether the hypernorm seems to be applied or not in the case in question. If the "potential" hypernorms fail the "reality check", the reasons to the failure need to be figured out and inquired into; *occasionally even the behaviour in question needs to be modified.* In some cases, if the hypernorm fails the "reality check", another candidate for a hypernorm needs to be identified or then alternatively we could change our attitudes and values conforming more adequately to our unethical behaviour, but not necessarily in the case of human dignity.

Some examples of needed authentic cases have been mentioned here, but more need to be identified through a collaborative effort. Examples of authentic ethical cases can be derived from the Vietnam War (e.g. case of My Lai), from "Operation Iraqi Freedom" but also from international "case-banks" (Lagacé-Roy 2007; Canada 2006). Unfortunately these kinds of "case-collections" pose only dilemmas and paradoxes and "universal right answers" cannot be delivered. The future will tell if we have any other option than to closely cooperate by allowing our national military practices to interact with the global ethical theories while the shared aim is to enable all of us to climb to qualitatively higher levels of moral development.

References

Annen, H. (2007). Military Pedagogy and Ethics: Current Challenges in Switzerland. In Toiskallio, J. (Eds.), *Ethical Education in the Military: What, How and Why in the 21st Century?* Helsinki: Edita Prima.

Ansari, A. (2006). *Confronting Iran: The Failure of American Foreign Policy and the Next Great Conflict in the Middle East.* New York: Basic Books.

Attfield, R. (2003). *Environmental Ethics: An Overview for the Twenty-First Century.* UK: Polity Press.

Bologna Process (2007). [On-line]. Available: http://ec.europa.eu/education/policies/educ/bologna/bologna_en.html.

Boatright, J.R. (2000). Contract Theory and Business Ethics: A Review of Ties That Bind. *Business and Society Review*, 105 (4), 452-466.

Calton, J.M. (2006). Social Contracting in a Pluralist Process of Moral Sense Making: A Dialogic Twist on the ISCT. *Journal of Business Ethics*, 68, 329-346.

Canada (2006). *Ethics in the Canadian Force: Making Tough Choices*. [On-line]. Available: www.cda.forces.gc.ca/cfli.

Carpenter, T.G., Bandow, D. (2004). *The Korean Conundrum: America's Troubled Relations with North and South Korea*. US: Palgrave Macmillan.

Cohen, S. & Libel, T. (2007). Training Military Personnel to Take Moral Decisions in Situations of Ethical Asymmetry. Lessons from the Israeli Experience in Toiskallio, J. (Eds.), *Ethical Education in the Military: What, How and Why in the 21st Century?* Helsinki: Edita Prima.

Colby, A. & Kohlberg, L. (1987). *The measurement of moral judgement.. Theoretical Foundations and Research Validation*. Volume 1. New York: Cambridge University Press.

Cussins, A. (1992). Content, Embodiment and Objectivity: The Theory of Cognitive Trails. *Mind*, 101, 651-688.

Cussins, A. (1993). Nonconceptual Content and the Elimination of Misconceived Composites! *Mind & Language*, 8 (2), 234-252.

Cussins, A. (2003). Postscript: Experience, Thought, and Activity in Gunther, Y.H. (Eds.), *Essays on Nonconceptual Content*. Cambridge, Massachusetts: The MIT Press.

Dershowitz, A. (2006). *Preemption: A Knife that Cuts Both Ways*. US: W.W. Norton & Company.

Donaldson, T., Dunfee, T.W. (1994). Toward a unified conception of business ethics: integrative social contracts theory. *Academy of Management Review*, 19 (2), 252-284.

Donaldson, T., Dunfee, T.W (1995). Integrative Social Contracts Theory: A Communitarian Conception of Economic Ethics. *Economics and Philosophy*, 11, 85-112.

Donaldson, T., Dunfee, T.W. (1999). *Ties That Bind: A Social Contracts Approach to Business Ethics*. US: Harvard Business School Press.

Donaldson, T., Dunfee, T.W. (2000). Securing the Ties that Bind: A Response to Commentators. *Business and Society Review*, 105 (4), 480-492.

Donaldson, T., Dunfee T.W. (2001). Precis for: Ties That Bind. *Business and Society Review*, 105 (4), 436-443.

Donaldson, T., Dunfee T.W. (2002). Ties that bind in business ethics: Social contracts and why they matter. *Journal of Banking & Finance*, 26, 1853-1865.

Dunfee, T.W. (2006). A Critical Perspective of Integrative Social Contracts Theory: Recurring Criticism and Next Generation Research Topics. *Journal of Business Ethics*, 68, 303-328.

Engeström, Y. (1987). *Learning by expanding: an activity -theoretical approach to developmental research*. Helsinki: Orienta-Konsultit.

Ercil, Y., Varoglu, K. (2002). Mental models and learning processes. In H. Florian (Eds.). *Military Pedagogy – an International Survey*. Frankfurt am Main: Peter Lang.

Fleck, G.(2005). Teaching Ethics: A Psychological Perspective in Micewski, E.R., Annen, H. (Ed.) (2005) *Military Ethics in Professional Military Education – Revisited*. Frankfurt am Main: Peter Lang.

Florian, H. (Eds.) (2002). *Military Pedagogy – an International Survey*. Frankfurt am Main: Peter Lang.

Foot, P. (2003). Challenges to Security and Defence Related Education and Training. In Toiskallio, J. (Ed.). *Identity, Ethics, and Soldiership*. Helsinki: Edita.

Fort, T.L. (2000). A Review of Donaldson and Dunfee's Ties that Bind: A Social Contracts Approach to Business Ethics. *Journal of Business Ethics*, 28, 383-387.

Ghoshal, S. (2005). Bad Management Theories Are Destroying Good Management Practices. *Academy of Management Learning & Education*, 4 (1), 75-91.

Gibbons, M., Limoges, C., Nowotny, H., Schwartzman, S., Scott, P. & Trow, M. (1994). *The new production of knowledge: the dynamics of science and research in contemporary societies*. UK: Sage publications.

Giddens, A. (1984). *The Constitution of Society. Outline of the Theory of Structuration*. Berkeley, Los Angeles: University of California Press.

Giddens, A. (1990). *The Consequences of Modernity*. Cambridge, U.K.: Polity Press.

Giddens, A. (2001). *Sociology. Fourth Edition*. Milan: Polity Press.

Habermas, J. (1984). *The Theory of Communicative Action: Reason and the Rationalization of Society*. Translated by Thomas McCarthy. Boston: Beacon Press.

Hartman, L.P., Shaw, B., Stevenson, R. (2003). Exploring the ethics and economics of global labor standards: a challenge to integrated social contract theory. *Business Ethics Quarterly*, 13 (2), 193-220.

Heinonen, R.E. (2007). Competence for Dialogue and Peace Building: Some rules and recommendations for intercultural encounters. In J. Toiskallio(Eds.), *Ethical Education in the Military: What, How and Why in the 21st Century?* Helsinki: Edita Prima.

Henk, D. (2005). Human Security: Relevance and Implications. *Parameters*, Summer, 91-106.

Henk, D. (2007). Human Security and the Military in the 21st Century. In J. Toiskallio (Eds.), *Ethical Education in the Military: What, How and Why in the 21st Century?* Helsinki: Edita Prima.

Heugens, P.P.M.A.R., Kaptein, M., van Oosterhout, J. (2006). The Ethics of the Node versus the Ethics of the Dyad? Reconciling Virtue Ethics and Contractualism. *Organization Studies*, 27 (3), 391-411.

Hofstede, G., Hofstede, G.J. (2005). *Cultures and Organizations: Software of the Mind. Revised and Expanded 2nd Edition.* US: McGraw-Hill.

Kasher, A., Yadlin, A. (2005). Military Ethics of Fighting Terror: An Israeli Perspectives. *Journal of Military Ethics*, 4 (1), 3-32.

Kohlberg, L. (1984). *Essays on Moral Development: The Psychology of Moral Development – The Nature and Validity of Moral Stages*. Volume II. San Francisco: Harper & Row, publishers.

Kuhn, T.S. (1962: 1970). *The Structure of Scientific Revolutions. Third Edition with a Postscript (1970)*. Chicago: The University of Chicago Press.

Lagacé-Roy, D. (2007). Military Ethics: A Case Study Method. In J. Toiskallio (Eds.), *Ethical Education in the Military: What, How and Why in the 21st Century?* Helsinki: Edita Prima.

Lakatos, I. (1978). *The methodology of scientific research programmes. Philosophical Papers*, Vol 1. Edited by John Worrall and Gregory Currie. Cambridge, UK: Cambridge University Press.

Lakatos, I. & Musgrave, A. (Eds.) (1970). *Criticism and the Growth of Knowledge*. US: Cambridge University Press.

Macgregor, D.A. (2003). *Transformation Under Fire: Revolutionizing How America Fights*. Westport, Connecticut: Praeger Publishers.

MacIntyre, A.(1985). *After Virtue: A Study in Moral Theory*. Second Edition. UK: Duckworth.

MacIntyre, A. (1998). *A Short History of Ethics*. Second Edition. UK: Routledge & Kegan Paul.

McCormack, G. (2004). *Target North Korea: Pushing North Korea to the Brink of Nuclear Catastrophe*. US: Publishers Group West.

Micewski, E.R. (Ed.). (2003). *Civil-Military Aspects of Military Ethics. Volume 1.* Austria: National Defense Academy Printing Office.

Micewski, E.R. (2005). On the Philosophical Framework for the Ethical Debate in Professional Military Education in Micewski, E.R., Annen, H. (Ed.) (2005) *Military Ethics in Professional Military Education – Revisited.* Germany: Peter Lang.

Micewski, E.R., Annen, H. (Ed.) (2005) *Military Ethics in Professional Military Education – Revisited.* Germany: Peter Lang.

Mutanen, A. (2007). Deliberation – Action – Responsibility: Philosophical Aspects of Professions and Soldiership. In J. Toiskallio (Eds.), *Ethical Education in the Military: What, How and Why in the 21st Century?* Helsinki: Edita Prima.

Mäkinen, J. (2005). Military education in the age of the Bologna Process. *Science and Weapon 2005*. (Tiede ja Ase 2005) 161-175. Jyväskylä: Gummerus Kirjapaino.

Mäkinen, J. (2006). *The Learning and Knowledge Creating School: Case of the Finnish National Defence College.* Ph.D. diss., the Japan Advanced Institute of Science and Technology [On-line]. Available: www.jaist.ac.jp/library/thesis/ks-doctor-2006/paper/jmakinen/paper.pdf.

Mäkinen, K. (2005). *Strategic Security: A Constructivist Investigation of Critical Security and Strategic Organisational Learning Issues: Towards a Theory of Security Development*. Helsinki: Edita Prima.

Nissinen, V. (2007). The Ethical Change of Military Command: the transformation of future military leadership education. In J. Toiskallio (Eds.), *Ethical Education in the Military: What, How and Why in the 21st Century?* Helsinki: Edita Prima.

Nowotny, H., Scott, P., Gibbons, M. (2001). *Re-Thinking Science: Knowledge and the Public in an Age of Uncertainty*. UK: Polity.

Pangle, T., Ahrensdorf, P. (1999). *Justice Among Nations: On the Moral Basis of Power and Peace*. US: University Press of Kansas.

Perry, D.L. (2005a). How ethics is taught at the U.S. Army War College. In E.R. Micewski & H. Annen (Eds.) *Military Ethics in Professional Military Education – Revisited*. Frankfurt am Main: Peter Lang.

Perry, D.L. (2005b). Commentary: Ambiguities in the "War on Terror". *Journal of Military Ethics*, 4 (1), 44-51.

Pitt, W, Ritter, S. (2002). *War On Iraq – What Team Bush Doesn't Want You to Know*. US: Context books.

Raviv, A. (2007). Command as the Profession of Military Officers: Concepts, Components, Education and Training Principles. In J. Toiskallio (Ed.), *Ethical Education in the Military: What, How and Why in the 21st Century?* Helsinki: Edita Prima.

Royl, W. (2002). Euro-Military Identity in Transition in Toiskallio, J., Royl, W., Heinonen, R.E., Halonen, P. (Eds.), *Cultures, Values and Future Soldiers*. Helsinki: Finnish National Defence College.

Royl, W. (2004). Euro-Military Identity as Core Element of Soldierly Spirit. In J. Toiskallio (Ed.), *Identity, Ethics, and Soldiership*. Helsinki: Edita.

Royl, W. (forthcoming a). *On the mutual compatibility of obedience and comradeship*. A paper presented at the 7th International Conference of Military Pedagogy in Bucharest 2006.

Royl, W. (forthcoming b). *The Universal Code of Ethics and its Mind-Leading Function in Military Conflicts?* A paper made for the 8th International Conference of Military Pedagogy in Jerusalem 2007.

Rowan, J.R. (2001). How binding the ties? Business ethics as integrative social contracts. *Business Ethics Quarterly*, 11 (2), 379-390.

Solomon, R.C. (1992: 1993). *Ethics and excellence: cooperation and integrity in business*. New York: Oxford University Press.

Stadler, C. (2003). Military Ethics as Part of a General System of Ethics. In E.R. Micewski (Ed.). *Civil-Military Aspects of Military Ethics.* Volume 1. Austria: National Defense Academy Printing Office.

Talerud, B. (2007). Ethos in and Ethics for Today's Armed Forces in the Western Societies: Why bother about military ethics and ethos? – Setting the stage. In J. Toiskallio(Ed.), *Ethical Education in the Military: What, How and Why in the 21st Century?* Helsinki: Edita Prima.

Toiskallio, J. (2003). Ethics, Military Pedagogy, and Action Competence. In E.R. Micecwski (Ed.), *Civil-Military Aspects of Military Ethics.* Volume 1. Austria: National Defense Academy Printing Office.

Toiskallio, J. (2004). *Identity, Ethics, and Soldiership.* Helsinki: Edita.

Toiskallio J. (2007). *Ethical Education in the Military: What, How and Why in the 21st Century?* Helsinki: Edita Prima.

Toiskallio, J., Royl, W., Heinonen, R.E. & Halonen, P. (2002). *Cultures, Values and Future Soldiers*. Helsinki: Department of Education of the Finnish National Defence College.

Toulmin, S. (1990). *Cosmopolis: Hidden Agenda of Modernity*. US: The University of Chicago Press.

Verweij, D. (2007). Military Ethics: A Contradiction in Terms? In J. Toiskallio (Ed.), *Ethical Education in the Military: What, How and Why in the 21st Century?* Helsinki: Edita Prima.

Värri, V-M. (2007). Some Problems of Ethics in Military Education: The question of ethics in the military space. In J. Toiskallio (Ed.), *Ethical Education in the Military: What, How and Why in the 21st Century?* Helsinki: Edita Prima.

Walzer, M. (2000). *Just and Unjust Wars: A Moral Argument with Historical Illustrations*. Third Edition. New York: Basic Books.

Wenger, E. (1998). *Communities of Practice: Learning, Meaning, and Identity.* Cambridge: Cambridge University Press.

Winer, S.(2004). *Between the Lies: Rise of the Media-Military-Industrial Complex.* UK: Southern University Press.

Wolfgang Royl

The Universal Code of Ethics and its Mind-Leading Function in Military Conflicts?

A foreigner who, while staying in Israel as a guest, ventures his opinion about ethical problems caused by the use of military violence in this country should exercise great restraint in doing so. The people who live here, as soldiers or civilians, have experienced years of nerve-wrecking low-intensity conflict, and they will not expect well-meant lectures to be of any help in their precarious situation. This is because there is just too big a discrepancy between the risk to life and limb they experience on a daily basis and any attempt to alleviate such a situation by means of military pedagogic analyses or well-intentioned advice. The philosophers of antiquity called this an aporia, which means a conflict situation, which leaves no way out. This is what turns drama into tragedy. Every day, plenty of instances occur on either side of a border wall as shells detonate in the territory of the respective enemy, destroying houses and killing people. And indeed, the suicide bomber is caught up in an aporia himself, as he believes that killing other people is an act pleasing to God. Kasher & Yadlin (2005) have tried to answer the difficult question as to what extent it might be legitimate and ethically justified to employ the military for the deliberate and preemptive killing of terrorists. This is the armed forces' response to a desperate situation, a response triggered by the terrorists who have set the scene for that situation. In doing so, they commit crimes against humanity. They mislead children and adolescents, and they misuse them by having them throw stones in confrontation with a military force. If the military resorts to the use of armed force to protect its members in response to such provocation, dead children and injured youths will then be presented to a worldwide audience as the telltale signs of inhuman militarism. Uninvolved observers of this conflict situation are rendered helpless by this. They may want to appeal to the people of good will of both conflict parties to cease these attacks and counterattacks of attrition and to peacefully come to an agreement as to claims and compensations. Attempts to achieve this on a small-step basis are made by means of international committees and UN conventions.

Military pedagogic deliberations should be situated within the sphere of such efforts at mediation, and especially those made by the 'International Association of Military Pedagogy', which seeks to regard military conflicts from an angle of global co-responsibility for world peace. What is crucial here is the spark, which may cause the hotbeds of tension within the global spectrum of conflicts to explode.

How could we use critical incidents that happen everywhere in the world to open the minds of soldiers so as to make them think and act along the lines of military pedagogy?

The Critical Incident Technique

John C. Flanagan (1954), a U.S. Air Force military psychologist, used this method during World War II for the purpose of diagnosing and analyzing problems encountered in the training of combat aircraft pilots. What had caused him to do so were incidents, which had occurred during training or mission assignments and had had particularly negative or positive effects on expected military performance. The method is strikingly simple in that it documents and analyzes particularly those incidents that interfere with the training sequence by causing a grave behavioral lapse, and in that it systematically studies instances of outstanding performance from which general derivations can be made as to the positive input that could be provided to shape the individual's behavior during training and mission assignments. This required a systematic sequence of diagnoses and analyses. The expert involved in the studies were then available to advise military units as to their findings. In this way, anybody who as an instructor needs to discharge his responsibilities within a fixed and prescribed perspective both in terms of time and training goals is given stress-reducing and success-oriented external support in his daily training routines. Indeed, such a task force-based qualification of instructors is not only suited for the training and missions of combat aircraft pilots and ground personnel. Instructive examples are gleaned from the sequences of events that occur during Army recruit training, and from the actions taken by soldiers, for example, during PfP missions, which are to be used for illustrative purposes in preparation for crisis situations.

Situation-related findings – when viewed from a scholarly perspective – require a frame of reference so that they may be arranged in order of importance and value. Military pedagogues may be expected to provide such order. Their expert knowledge and their capabilities in conducting research in the field of military pedagogy should therefore make them obliged to use the findings obtained through the analysis and treatment of critical incidents in the military for the purpose of forming a theory guided by practical considerations and for conducting practical work guided by theoretical considerations.

May Kant's Categorical Imperative Claim to be of 'Universal Validity' Even in Military Conflicts?

This question relates to Kant's deliberations as to the possibility of a generally applicable moral norm, with 'generally applicable' meaning: to be of significance to all human beings in this world in general and to each one of them in particular. Kant derives the claim to universal validity from the individual's ability to apply reason. Following Flanagan's considerations, one would be able – when looking back on Kant's philosophy – to discern a link to the point which is at issue here, namely the question as to whether there are universalistic ethics which would apply to military actions, too. As it is, in the history of philosophy the issue has been placed within the methodical approach of 'criticism' and, in its own terms, is a major intellectual achievement, which received, and still receives today, attention from all over the world.

One may not generally assume that the political actors who are responsible for the ongoing situations of political conflict do have knowledge of Immanuel Kant's deliberations on determining moral responsibility. One may have to allow for the possibility that political perpetrators do not give a damn about the sharp-mindedly phrased moral requirements conclusively set forth by the philosopher from Königsberg. However, it should not be concluded therefrom that philosophical bodies of thought like the one we owe to Kant are devoid of practical relevance, aptly suited for being pondered upon at the universities but not for anything else. So as not to turn this restriction into an impediment for the ensuing statements, recourse is made to a methodical detail of the philosophy according to John Rawls (1971). His major work *A Theory of Justice* proceeds on what Kant had formulated some 200 years before. Rawls applies an experiment of thought. Any counterarguments, which might come up and interfere with the development and presentation of ideas about the main issue of 'justice as fairness', are, for the time being, relegated to the background by the 'veil of ignorance' (Rawls, 159ff). In this way, it is initially of no significance whether members of Hamas, al-Qaida fighters, or members of the al-Aksa brigades consider ethics-based morals to be the work of the devil because these morals are taught at the universities of the Occident and are not to be found in the Koran.

Groundwork on the Metaphysics of Morals

In his "Groundwork on the Metaphysics of Morals", Kant develops the concept of 'duty' in combination with that of 'morals'. This is of significance insofar as the so-called 'categorical imperative' for acting ethically is derived from the supreme principle of good will. When applied to ethical actions, 'principle' means that the individual, in adhering to it, should act out of 'duty' rather than 'inclination'. According to Kant "duty is the necessity to act out of reverence for the law". By linking the 'duty' to act ethically to the claim of adhering to the law of morals without any restrictions, all possible limiting conditions are excluded. Statements such as "maybe, yes" or "depends on the circumstances", which would put a limit to the obligation of acting ethically, are ruled out from the start. Instead, behavior is to be considered as moral if it is suited for conversion into a generally applicable law. Thus, the claim of acting morally becomes unconditional and therefore generally applicable.

Applied to military conflicts, this means that civilians must be spared as a matter of general principle, and not just because troops have run out of ammunition or because a UN observer happens to be nearby. This categorical demand is made by the soldier on himself if he knows that he is personally responsible for his own actions. This is true both for military superiors and for their subordinates. What needs to be determined, then, is whether such a moral component that goes beyond knowing the legal provisions as to the protection of cultural assets or the treatment of prisoners has already found its place in military training or whether it should be accorded such a place. If an instructor seeks to enhance a recruit's fighting spirit by encouraging him to scream 'motherfucker' with every burst of fire so as to debase his adversary as an individual, then the dulling of the mind is regarded as a prerequisite to being ready to fight. On the other extreme, there is the World War I example of Langemarck, a Belgian village where on 10 November 1914 German soldiers suffered heavy casualties when they stormed uphill against British machine gun nests in skirmish line formation, singing the national anthem with the enthusiasm instilled by youth. The losses on the German side – which were the result of a command error – were reported to be 2,000 men on a single day.

Refusing an Order for Reasons of Conscience

From among the critical incidents which could be put in relation to the 'categorical imperative' as regards thinking and acting along the lines military pedagogy, one incident is selected here which deals with the reported behavior of Israeli aircraft crews. A second critical incident to be discussed is the order that was refused by a major serving in the German *Bundeswehr*.

1. The bombing of refugee camps

In a conversation with Dan Tamir, a member of the Israeli peace movement, Elke Durak (2002) discussed why it is the case that actually there are people in the

Israeli Army who refuse to obey orders. Tamir refers to these soldiers as true patriots. In the face of Palestinian suicide bombing attacks, though, soldiers who refuse orders are considered to be traitors to the people of Israel.

It can be discerned from the interview that while these peace activists support the defense of Israel, they do not want to participate in military actions, which in their eyes constitute a violation of human rights. Etgar Keret (2004) deals with such violations, too. He criticizes Israeli guard personnel who, against the established procedure of military ethics, subject Palestinians to chicanery when they want to pass through the Haware checkpoint established near Nablus.

Those who refuse orders in the Israeli Army accept that they will be sent to jail. According to published reports, so did 27 Israeli combat pilots who refused to fly attacks against Palestinian refugee camps because the victims of these attacks were mostly civilians.

The example of these bomber pilots in particular serves to highlight the dilemma, which political leaders – and thus the military as well – find themselves caught up in. This dilemma is not just about civilians being injured or killed by these air attacks. Moral criticism is also directed against those on the Palestinian side who misuse the civilian population as a shield. Those who launch missiles against Israeli settlements and cities out of Palestinian residential areas deliberately accept the death of noncombatants in case of Israeli retaliation. What is perfidious about this kind of warfare is the fact that on one hand the world public expresses its concern for the victims among the Palestinian civilian population, while on the other it was Hamas and other terrorist groups that intentionally caused these people to become victims in the first place. Although this strategy is sufficiently known by now, pictorial coverage of the devastating effects which Israeli counterstrikes have on the Palestinian civilian population cause the world's public opinion to side against Israel.

Another dilemma is revealed by the fact that Israel's highly armed forces have to deal with guerrilla fighters who, in a manner well proven throughout the history of war, imitate tactics that turned out to be successful in other political conflicts, as in China, for example. Fighters like these – partisans, guerrillas, irregular forces or insurgents – have often proved superior to conventional forces in terms of mobility, concealment and fighting morale.

Therefore, the dilemma is threefold:

a) The civilian population is made a target of attack.
b) A regular army using massive retaliation against guerrilla fighters will cause the opposing side's affected civilian population and the world public to turn against it.

c) In an asymmetric war, it is difficult for 'Goliath' to achieve military success against an actually weaker, hard-to-find opponent and his propaganda.

Both the members of the Israeli peace movement and the soldiers who refuse orders which they think require them to commit violations of human rights are putting moral-psychological dilemmas at issue. Dealing with this is a function of military pedagogy in the training of officers and enlisted personnel. As long as moral considerations are not avoided in the discussions of critical incidents, there is a chance that negative attitudes towards fighting activities may be diminished and that positive components of behavior within one's own army and towards the opponent may be reinforced. The results of such efforts should be reported in the form of differentiated casuistics.

2. The case of Major Florian Pfaff, an officer in the German Bundeswehr

Because of negative lessons learned in German military history, the right of conscientious objection was laid down in the Basic Law after the founding of the Federal Republic of Germany. It is not only draftees who may refuse to take part in war missions; officers and enlisted personnel may do so, too. In practice, this provision is rarely, if ever, made use of by people who already serve in the military. The public's attention is aroused all the more, then, by isolated cases in which individuals claim their constitutional right to conscientious objection. This attention becomes even more heightened if an officer succeeds in securing this right even before the Federal Administrative Court, which is the highest legal body competent in these matters.

Major Pfaff, who is an expert translating management tasks in the field of materiel organization (fuel, bombs, etc.) into IT programs, was employed as a section chief to develop a complex logistics program. The pertinent software was to coordinate individual measures taken separately until then so as to render them more efficient. Mr. Pfaff held the opinion – which surely was not unfounded – that such an integrated logistics program could, among other applications, also be helpful in coordinating military supplies in Iraq, and might therefore be used in that country. Because the second Iraqi war and the subsequent occupation of the country by allied forces had occurred without legal foundation – as there had been no resolution by the UN Security Council – Mj. Pfaff could not reconcile with his conscience the development of such an IT program which might be used to support the USA, which had led a war of aggression against Iraq that was not in keeping with international law and which maintained troops in the country as an occupying force. The court accepted Mj. Pfaff's opinion and acquitted him on 21 June 2005, also rescinding his demotion to captain imposed because of his refusal of orders.

What is remarkable about this court ruling is that it upheld the basic right of freedom of conscience provided for in article 4 paragraphs 3 of the Basic Law, which codifies the right of conscientious objection. In the opinion given by the court in its remarkable decision, the court held that the term 'military service' even included the rendering of a service that might be used in a war that was not justified under international law. Refusing to render such a service – which is far remote from combatant service – was legally recognized as a decision dictated by conscience. The court accepted that the accused officer's refusal of orders was subjectively justified.

Did this soldier act according to the categorical imperative, which says: "Act only according to that maxim whereby you can at the same time will that it should become a universal law?"

Undoubtedly, the court ruling in favor of Major Pfaff is a critical incident. It puts at issue a consistency to which superior officers not always react appropriately who attempt to swiftly take counteractive measures out of concern that an individual's refusal to obey might undermine the morale of a whole unit.

Now, should the major's immediate superiors be blamed for underestimating his ethical rigorism? Should one expect from them to have arrived at a similar conclusion dictated by conscience, which the major had arrived at? Would the problem have arisen at all if a similarly competent IT expert had been found who might not have that deeply engaged himself in thoughts about the 'good' and 'evil' of his doings? Without insinuating answers to these questions, the case of Major Pfaff challenges all those who have positions of responsibility, from the immediate superiors to the commanding general, to take intelligent and morally sensitive subordinates seriously. This means to subject one's own person to a moral decision process and to heighten one's awareness of what is right and what is not, a process of which even those may profit who are less smart and less morally sensitive.

After succeeding with his case before the Federal Administrative Court, the major has not returned to military business as usual. Meanwhile, he is being courted by institutions, which take a critical stance towards the government and the military and which use him as a figurehead with the intent to shed a negative light on the armed forces. This behavior is no longer in keeping with a maxim whereby you can at the same time will that it should become a universal law.

Moral Dealings with Guilt in Past and Present Times

Critical incidents as defined by Flanagan invariably will include grounds for moral conflict, as is shown by the examples of the bombing of Palestinian refugee camps and by the case of the *Bundeswehr* major.

Two critical incidents of global format are to be cited now which highlight the moral problem linked to the two narrowly defined examples of disobedience just discussed. One is about the collective crime committed in the former Soviet empire; the other one is about the ongoing disregard for human rights in southern Sudan. Both incidents corroborate the plurality of collective crime. They are a downright challenge for people to deliberate on how military ethics should be implemented.

Archipelago Gulag

Aleksandr Solzhenitsyn demonstrated impressively how Stalin wrought havoc on his own population. The examples of political terror used by Stalin to maintain his rule speak for themselves. It would appear that the description of Soviet internment camps and the memories of crimes against humanity committed during Stalin's regime through random arrests and random verdicts against presumed opponents of the regime have already been relegated to the grave of history. So far, there has been no debate on Stalin's rule of terror similar to the discussion that was, and still is, conducted in Germany on Nazi dictatorship. Even if dealing with political crimes will be followed by claims for recompensation by victims and their families, the costs are far outweighed by the moral gains, which are provided by the fact that the potentially ensuing cost of a dictatorship and the crimes against humanity committed under it contain a high potential of remembrance. In analogy to the individual rule of behavior: Crime doesn't pay! The collective responsibility for crimes committed by the state – especially when it is justified and called upon under international law, as was the case with the war crime trials in The Hague – is an important contribution to the global effectiveness of moral standards. Solzhenitsyn established something akin to a case record, documenting the people involved – perpetrators and victims – as well as their actions and environment. In this way, Solzhenitsyn established publicity.

As is generally known, there has been no critical dealing with Stalinist crimes under legal aspects, nor have victims been recompensed. This is disappointing for the public's sense of justice and for those affected by terror and violence. Evil deeds and suffering are not forgotten, though, as the criminal proceedings are initiated before the court of history. It remains a valid demand that the community of nations should bring the perpetrators of collective crimes to justice and mete out adequate punishment. As long as there are crimes against humanity that have not been atoned for, scholars have the task of keeping up moral awareness. Every sentence written to make sure that those facts are not forgotten, and every argument brought forward to identify the perpetrators and the victims is like a brief that goes on the 'prosecution of crimes against humanity' file. And a comprehensive file it is. It is made even more so by perpetrators remaining undiscovered because of the opposing activities of investigators and those attempting to hush things up, or because perpetrators remain beyond reach in a neutralizing political situation. As long as perpetrators maintain their political power to protect themselves against public opinion, against the threat of punishment, and against criminal prosecution, they can continue to erect monument after monument for their own adolation, to put people in awe, and to demonstrate their own power. At the same time, however, such monuments are a contemporary testimony to these crimes.

Genocide in Darfur

A particular challenge to military ethics was published in the form of an impressive and credible research work by Gérard Prunier (2006). His research draws a picture of the politico-military development in Darfur from 1916 to 2005. This region in the west of Sudan comes under the jurisdiction of the Sudan national government, which resides in the capital of Khartoum. It is from there that the people living in the three provinces of Darfur should be afforded protection. Obviously they are not, for reasons of domestic policy.

Social order in Darfur is represented by the tribal community, a form of society shaped among the nomads in the desert climate of these regions. The pastoral people are organized in clans and roam the savanna with their cattle herds, searching for adequate pasture. Over the decades, though, they have in part adopted a peasant culture and formed village communities.

Actually, it does not come as a surprise that chaos and arbitrariness rule in such a region, which has more or less been sacrificed by its own government to the political interests of the adjacent states of Egypt, Libya and Chad. In this process, the cultural difference between the ethnic group of Africans and the ethnic group of Arabs is defined along racist lines, which in turn provides an excuse for justifying military transgressions against the population of Darfur. According to Prunier's report, rights of ownership are disregarded in the most blatant form. As human life has no value in the eyes of the attackers and as people have only insufficient means to defend themselves, Africans living in the region are killed and robbed in mounted attacks conducted by Arab mercenary troops called Janjaweed. Girls and women are first raped and then killed by the attackers. The destructive scenario in the region is aggravated by periods of draught, and in his current assessment of the situation in Darfur, Prunier diagnoses a 'genocide', which only for formal reasons he refers to as 'ambiguous', with the 'ambiguous' part about it being just the fact that the perpetrators have not been specifically identified. What is lacking is something like Adolf Hitler's *Mein Kampf*, i.e. a book in which one could read in black on white about the intention to annihilate the people of Darfur, based on which the perpetrators and their clients could be indicted before the world public.

While the responsible government in Khartoum is entangled in corruption, leaving the people of Darfur without protection against the terror of the Arab mounted militia, one author has the courage to name names as regards those who are involved in the disaster and the UN bodies whose jurisdiction would include such conflicts. He blames them for the genocide-like destruction of human lives by military attacks, denouncing the insidious destruction of the native population by well poisoning and the deliberate killing of donkeys, which are an indispensable means of transport in the region. The destruction of humans is accelerated by a lack of sanitary facilities and a medical service, which is even less than insufficient. It fits the picture that the mounted militiamen generally will kill girls and women after raping them: Unintended population growth among the African population as a result of the rapes is to be avoided.

'Genocide' is an apt word for this form of depriving the local people of their means of living and reducing their numbers by way of murder. Because of current media coverage, the world public has sufficient knowledge of the Arab militia's reign of terror against the local population, and of the militia's clients. The UN crisis staff knows about the situation in Darfur. The UN Secretary General voiced his complaints in Khartoum. But the murders in Darfur in the form of ethnic cleansing continue under the eyes of the world public. Meanwhile, the International Court of Justice at The Hague has appointed one Luis Moreno-Ocampo as chief prosecutor for the Darfur case, after the UN World Security Council issued a resolution on 1 April 2005 denouncing the crimes committed by the government in Khartoum.

World Ethos

The Swiss theologian Hans Küng earned great merits for advocating the World Ethos project (1992) and for demanding that human rights should be enhanced and put into more concrete terms by means of a joint fundamental ethos (p. 87). As well founded as Küng's subsequent rationale for it might have been, such a sensible goal is bound to be faced by institutions which seek to maintain religious systems and their positions within them with the intention to claim sole representation. The idea of becoming actively involved in a world religion is a cause of concern for the dignitaries representing the various religious faiths. For jointly held values to be expressly included into the moral principles of action of the various denominations would require, for example, that they be brought into consonance with the various concepts of salvation held by the churches. The intellectual potential of the clergy would no doubt allow them to do that, as is demonstrated now and then in theory and in practice in their ecumenical work. A subsumption of clerical power is not to be foreseen, however. It may be discerned from public media coverage in the Western world, though, that something exists that might be called a world conscience. It reacts to news from crisis areas, and obviously it causes any calls for donations to be responded to not only by relief organizations but also by private individuals worldwide.

World Ethos and World Conscience

Relieved of the Catholic precept of obedience, religious scholar Hans Küng encouraged a worldwide discussion on joint ethical values and modes of behavior to be used as guidelines for all people. This is about reaching an agreement on the concepts of values and principles of moral behavior, which already are held in common by the world religions. As early as 1893, a conference was held initiated by Charles Carroll Bonney, a lawyer from Chicago, which made a remarkable contribution to ecclesiastical history as the 'Parliament of World Religions'. The second conference of this type took place in 1993. The result of this gathering attended by 6,500 representatives of religious denominations was the adoption of the 'Declaration on World Ethos' subsequently published by Küng & Kuschel (1993).

Contained therein are the four so-called irrevocable directives:

1. Commitment to a Culture of Non-violence and Respect for Life
2. Commitment to a Culture of Solidarity and a Just Economic Order
3. Commitment to a Culture of Tolerance and a Life of Truthfulness
4. Commitment to a Culture of Equal Rights and Partnership between Men and Women

Concrete derivations to be made from these directives include:

To 1: Partnership for Peace missions are an example of fulfilling the commitment to non-violence, as they are designed to prevent violence and to bring peace, or a least a truce, to crisis regions. Whether efforts to mediate between antagonistic groups as part of such a mission will succeed will essentially depend on the soldiers' own attitude towards ethical principles and their willingness to follow moral codes in their personal actions. This creates a continuing task of military training and leadership.

To 2: By combining a world community of solidarity and a world economic order, the Parliament of World Religions has taken up a matter of central concern.

This is to put up a front against the economic diagnosis that the rich are getting richer while the poor are getting poorer. Reference is made to the necessity to provide for an economic balance, for example to benefit Third World countries. The industrialized nations are expected to share their wealth by raising the standard of living, by fighting corruption, and by providing for adequate wages in these countries. A good world conscience is not to be had for free. Since the correlation between poverty and a high birth rate is an established fact, one may expect the birth rate to decline – and this applies to the Palestinian regions as well – as the standard of living is raised and a safe existence is provided for those who have reached old age. Families with an adequate income normally have fewer children so as to be able to provide adequate school education and vocational training for them.

To 3: Culture of tolerance means to mutually dispense with inappropriate claims of rule. This applies both to the private relations between citizens, including those of different ethnic origins, and the relations between nations with different levels of economic performance. In comparison to earlier centuries, today's organization of the United Nations may be viewed as a body by means of which even military action may be taken in the interest of the community of nations if such action serves to restore peace.

The fact that Israel had to witness Soviet senior officers providing guidance to Egyptian troops (Luttwak & Horowitz, 1975) and, according to reports, Soviet officers sometimes would even fly combat sorties wearing Egyptian uniforms during the Six Days' War is indicative of the kind of tasks that still await worldwide implementation of military ethics.

To 4: Culture of equal rights means to undertake serious steps towards the implementation of human rights. This directly addresses the relation between the genders. It should be credibly represented in the families and in the occupational sphere, including the military officer's profession. A demand is made that all individuals should be invited to take this duty upon themselves and to categorically apply its substance in their actions. Indeed it should be considered whether a memorial should be built to the raped girls and women of Darfur and Kosovo, and also to the Israeli child and the Palestinian child killed in the Intifada, for them to represent the many victims who were robbed of their future.

Exemplary incidents such as are known to have occurred in Soviet internment camps or the racist motivated acts of war against the population of Darfur amply demonstrate the real and large-scale need for competency in moral behavior to be imparted. The 'Parliament of World Religions' formulated ethical directives in a pronounced setting. Such directives imply and emphasize also the necessity to make ethics an issue of teaching and of translation into moral action on a worldwide basis, including the military sphere. In this effort, one may rightfully expect to receive support from the 'International Association of Military Pedagogy', as repeated attempts can be made through the Association's members to inform the respective nations' armed forces and to provide scholarly expertise on the matter.

Military Ethics in the International Discourse

If, with respect to the question as to the scope of applicability of military ethics, one were to adopt a universalistic point of view in the sense of Kant, it would be of no relevance how many people fall victim to collective crime. If the individual human being is never to be a means to an end but an end in and of himself/herself, then a cattle herder in Darfur is no less an end in and of himself than was a Russian peasant and his family during the founding period of the Soviet Union. If, as is done, a devastating judgment is delivered as to the cleansing sprees conducted during Stalin's dictatorship, then the political subterfuge of government responsibles in Khartoum, Tripoli or the Republic of Chad must not be exempted. This is comprehensible not least because, then as now, the stigma of 'unworthy life' is used to justify the deeds of destruction.

As for the declaration of human rights in their universalistic diction and their inclusion in the constitutions of the free world: Are they not devalued and questioned by knowingly leaving ethnic groups that are unable to defend themselves against corrupt governments and genocidal assassins to their own devices and by half-heartedly deploring their plight? Is it not that so-called cultural nations cause themselves to definitely lose credibility if their governments keep the conflict situations in restless provinces going like so many open sores for the sake of international power games?

Since the military, with its authority to use force, has a direct need for its actions to be ethically regulated, there is an unrestricted justification to develop and treat military ethics – soldierly ethics – as a special and critical field of application. This is done in military training on a continuing basis with each new batch of recruits that comes in. Moral limits are set for the use of weapons both in combat and in self-defense situations. To accord combatant status to somebody means to oblige that person to adhere to the Geneva Conventions and the Hague Rules: The wounded will not be shot but taken care of; prisoners will be registered and, if possible, will be administered adequate care. If a prisoner dies while in captivity, his family will be informed as to his or her death, with the prisoner's personal effects to be delivered to the bereaved. The civilian population will not be intentionally exposed to combat action.

The intention behind these regulations shows criteria of the humanization of war, which should take priority over enforcing one's interests of power with military means. As to these accords, the question is raised whose task it actually is in a critical military situation to ensure that such orders and conventions are actually imposed. It would not be anything out of the ordinary if the first who would come to mind would be the officers, who bear responsibility for their soldiers. And it should be an ordinary matter no less if the application of universalistic ethics to military conflict situations were to be made a responsibility of the individual soldiers. For military ethics, as a moral attitude of the soldier, to take effect in combat action requires a corresponding moral awareness to be developed. This is why military pedagogues are called upon to contribute to the development of corresponding awareness facts and value attitudes as well as to their application in everyday military training.

In his deliberations on 'taskmaster war', Frankfurt-based political scientist Karl Otto Hondrich (1992) came to an all too sobering conclusion: *"There can be no society without learning. Any attempt at imagining a world without war needs to provide an answer to the question as to how world society may continue to learn without it*" (p. 9). Indeed, this question remains to be answered.

The same holds true for his second statement: "Individual learning changes the individual's concepts of himself or herself as related to other persons, while collective learning changes (the same individuals') concepts of their own society as related to other societies" (p. 45).

For an internationally framed military pedagogy, such lucid statements reinforce the necessity to take these approaches to 'world ethics' seriously and to arrange for the idea of obligatory participation in a world conscience to take effect in training and education. Such military pedagogy adopts a viewpoint that transcends national fixations, as purposeful and well founded by history as they may be. In a global perspective, the proper thing to achieve is imbuement with moral norms and values in the form of a world conscience to the benefit of a world superego, as it were, which would then be able to attain public awareness of good and bad military actions, and to assess them too. A helpful instrument – which is effective even now, as it is generally accessible – is the objective coverage of radio and television about what goes on in the world.

Literary contributions for which access has so far been limited mostly to a readership interested in military pedagogy include those written as part of an initiative started by the Finnish professor Jarmo Toiskallio, who teaches and conducts research at Defence University, Helsinki. Added to this, from an international viewpoint, is a main field of research on issues of military ethics, which Brigadier General Dr. Edwin Micewski, Institute for Human and Social Sciences at the National Defence Academy, Vienna, was able to develop further. Owing to the General's efforts, it was also possible to hold conventions in connection with the Austrian *Bundesheer*'s Civil-Military Cooperation Initiative which proved to be extremely fruitful from a scientific perspective. Following a suggestion of Professor Edmund van Trotsenburg, formerly with the University of Klagenfurt, the publishing house *Europäischer Verlag der Wissenschaften*, Peter Lang, Frankfurt/Main, has graciously agreed to provide its 'Studies for Military Pedagogy, Military Science & Security' series as a platform for contributions in the field of military pedagogy. In 2005, work results titled 'Military Ethics in Professional Military Education' were published in Vol. 9 of the series by Edwin Micewski, National Defense Academy, Vienna, and Hubert Annen, Military Academy at the ETH Zurich (eds.). In 2007, Annen and Royl published selected contributions of fellow scholars on the subject of 'Military Pedagogy Progress' in Vol. 10 of the series. Moreover, the research findings of the international conventions of military pedagogues held at Bucharest in 2006 and at Jerusalem in 2007 were published in the present book.

Summary

One would do well not to overlook the question mark in the heading of this contribution: "The Universal Code of Ethics and its Mind-Leading Function in Military Conflicts?" It expresses the fact that, from an international point of consideration and assessment, the translation of universalistic ethics into military actions still has room for development. Wherever impediments may arise in this process, it remains to be hoped for that international leadership cadres – including those of the Middle East who, in no rare cases, have for themselves experienced free access to science – will also support the process of historical development. At these universities, they generally had mental access to information about the intellectual history of the Age of Enlightenment. To remain on a par with this is a task of the Western world, too, which may well benefit from the criticism voiced about its economic ways and its lifestyle. "Las Vegas is not a must-have!"

The way chosen was to use the critical incident technique for highlighting and assessing matters of concern which throw in relief the issues of military ethics.

In this way, the issues of military ethics directly lead to justness in dealing with problem cases in one's own armed forces, and they provide for an appropriate treatment of opponents. To the extent possible, it is avoided that uninvolved civilians are affected by wars or war-like activities.

A plea was made that one should follow the categorical imperative as was formulated and justified by Kant (1786), the more so since John Rawls (1971) further developed this categorical moral claim in our times into a well-founded 'justice as fairness'.

The question as to whether Kant's Categorical Imperative may, and should, be claimed to have 'universal validity' even in military conflicts was answered in the affirmative. Two specific conflict cases were selected so as to portray the moral dealing with guilt in past and present times. Two examples were given to provide insight into how moral judgment and behavior may be confirmed both in retrospect (Solzhenitsyn) and by a modern-day German court of law (case of Florian Pfaff).

An internationally framed military pedagogy will be expected to transpose ideas as to the national use of military ethics to the level of world ethics. The guidelines of such ethics have been formulated by the 'Parliament of World Religions', and these guidelines are of relevance to military pedagogy, too. They are relevant in that soldiers are to be familiarized in their training and education with ethical issues and the modes of moral behavior in military conflicts. This is done intending to afford soldiers the opportunity – by imbuing them with the principles of ethics – to bring their own behavior (on and off duty) in consonance with will and conscience. PfP missions in particular offer the chance to provide moral rationale to support solutions to conflicts among the local population or conflicts among UN force members. The national perspective of those involved should be able to be overruled by one that helps to open up a mental path to the necessity of 'world peace'. This seems appropriate especially because conflicts, which occur on location normally, take place within the context of international tensions and are thus governed by these tensions.

Finally, reference was made to the findings of the International Association of Military Pedagogy, which has convened since 1990 to provide a forum of exchange as to the progress already made and the steps considered as necessary in the field of military training and education. The results of these annual conventions have been published by Peter Lang, Frankfurt/Main and are thus available for further public discourse. This is an important continuing task, as is shown even now by the current convention held at the Jerusalem Centre for Ethics (2007).

References

Bundesverwaltungsgericht (2005). *Urteil im Disziplinarverfahren gegen Major F. Pf. wg. Befehlsverweigerung*. Leipzig: 2. Wehrdienstsenat.

Durak, E. (2002). *Befehlsverweigerung ist echter Patriotismus*. A conversation with Dan Tamir, Yesh Gvul peace movement, on 15 May 2002 broadcast by *Deutschlandfunk*.

Flanagan, J- C. (1954). The Critical Incident Technique. *Psychological Bulletin*, 51 (4), 327-358.

Hondrich, K.- O. (1992). *Lehrmeister Krieg*. Reinbek: Rowohlt.

Kant, I. (1786). *Grundlegung zur Metaphysik der Sitten*. Riga: J.F.Hartknoch.

Kasher, A. & Yadlin, A. (2005). Assassination and Preventive Killing. *SAIS Review*, 25 (1). 41-57.

Keret, E. (2004). *Ethischer Codex für die israelische Armee oder "Sprich Hebräisch oder halt›s Maul."* Znet Deutschland 11.8.2004.

Küng, H. (1990). *Projekt Weltethos*. München: Piper.

Küng, H. & Kuschel, K.-J. (1993). *Erklärung zum Weltethos. Die Deklaration des Parlaments der Weltreligionen (4.9.1993 in Chicago).* München: Piper.

Luttwak, E. & Horowitz, D. (1975). *The Israeli Army*. London: Allen Lane.

Prunier, G. (2006). *Darfur. Der 'uneindeutige' Genozid*. Hamburg: HIS-Verlags-gesellschaft.

Rawls, J. (1971). *A Theory of Justice*. Harvard: President of Harvard College.

Solschenizyn, A. (1974). *Der Archipel Gulag*. Bern: Scherz.

Hermann Jung

Situational Ethics and Cultural Awareness for Task Forces during Intervention-Stabilization-Transformation Operations (IST)[1]

A Working Definition

IST operations are sustained and integrated interagency (often multinational) activities to project power to an ungoverned area, failed state, state-in-conflict, or chronic aggressor state, to quickly restore order, and then to ameliorate the source of instability or aggression by transforming that state into a stable, progressive member of the international community. Planning for IST operations must be holistic. Still, of the three functions-intervention, stabilization, and transformation – each requires a different configuration of skill sets from the deployed military forces. The skills and organization needed for intervention are already resident in current forces and likely to be even more evident in the future. Transformed forces augment their expeditionary and joint mindset. Global strategic speed, information superiority, and the ability to undertake rapid decisive operations are all key. But for stabilization, the requirements shift in part because the psychological tasks necessary for the mission become significantly more complex. To succeed at intervention, the forces only need to show enough power against their opponents to end resistance.

Stability and Support operations demand for leaders, who can work with communities with widely divergent backgrounds. For instance the Provincial Reconstruction Teams (PRT) in Afghanistan under the command of the International Security Force (ISAF) are planning to assume security responsibility in support of the Afghan government with the US led Combined Forces Command (CFC-A). The mission of all PRT is to assist the Government of Afghanistan to extend its authority in order to facilitate the development of a stable and secure environment. To fulfill this mission you need a common understanding of the causes of instability. This requires intercultural competence. In an area of a weak government these teams are primarily engaged with non-governmental actors (friends and opponents) in a contest for the ungoverned space and this not only in the physical sense but also in the cognitive domain, in the minds of the population. We have to win wars of ideas in a complex world. Winning wars of ideas means to use all

1 This article is a protocol of a lecture given by the author during a pedagogical conference held in Jerusalem on June 2007

means toward a better situational understanding. And it needs a cultural literally educated soldier. That brings us a better cultural understanding.

Cultural Competency

Is about building successful teams in a culturally diverse environment with a common vision, and performing acceptable processes that benefit from cultural diversity. Communication and trust are often more important than activity. Military leaders are trained to make decisions rapidly with little time available for discussion, debate, or consideration of dissenting views. Events involving potential destruction or violence demand one minute managers or leaders, and doing so entails rapidly obtaining key facts and essential information, internal processing, and then choosing and implementing an appropriate course of action (COA).

Cultural literacy is about understanding your individual cultural patterns and knowing your own cultural norms. Understanding how your culture affects someone else's culture can profoundly affect any COA's chances for success. Military leaders have an additional challenge; they must understand and appreciate their own military culture, their nation's culture, and the operational area's culture. To effectively manage the dynamics of differences, leaders must learn effective strategies for solving conflict among diverse people and organizations. They must also understand how historic distrust affects current interactions, realizing that one might misjudge others' actions based on learned expectations. Cultural background is one of the primary sources of our self-definition, expression, and relationships within groups and communities. But when we experience a new cultural environment, we are likely to experience conflict between our own cultural predispositions and the values, beliefs, and opinions of the "host culture." Cultures often experience alterations in cultural identity, which might create significant insecurity in both interacting cultures, calling into question identity, and in values, which might result in an adversarial relationship. Culturally literate soldiers understand and appreciate their own beliefs, behaviors, values, and norms but they are also aware of how their perspectives might affect other cultures' views. Achieving self-awareness of our own cultural assumptions enables us to use this understanding in relations with others

Stabilization-Transformation

Means to deter potential opponents, win „fence sitters", convince external actors, and reassure military presence always carrying the threat of a quick forceful action up the sleeve.

During stabilization, potential opponents must be deterred or dissuaded, supporters of our forces need to be reassured, "fence sitters" won over, and external actors convinced to support the effort, or at least not interfere with our operations. Only a force capable of providing a large scale extended, culturally and politically sensitive presence, which builds campaigns around desired psychological effects, can attain such a complex array of psychological objectives. In some cases, the military presence may be friendly and reassuring, but the threat of a quick, forceful reaction must always be palpable. In other words, the most successful stabilization force is one that wears both the mailed gauntlet and the velvet glove. In addition, the military should work more closely with nonmilitary organizations, both governmental and nongovernmental, during stabilization. As a consequence, psychological components will gain even more in importance. Success at IST operations requires both proficiency and an integration of non-kinetic efforts even with non-military partners. Even though transformation may be the most important dimension of IST operations and the ultimate guarantor of strategic success, the direct role for the military is at its lowest. Creating an indigenous security force that is both effective and professional is key; it is here that contractors can make a valuable contribution to the effort. The military's major functions remain continued oversight and direction and of course the provision of security.

Capability Enhancement of Perceiving Cultural Differences

Cultural differences can only be perceived on the background of our cultural norms, including thoughts, behaviors and patterns of communication, customs, beliefs and institutions. Two such capabilities are intelligence and psychological operations. But while existing Psychological Operations Forces have impressive skills, they are primarily designed to support commanders in combat. Information and psychological activities are a major operational axis in IST, not a supporting one. This enhancement requires innovation. For instance, winning wars of ideas requires the sophisticated crafting of media reaching into a target society.

Rebuilding civil security forces is another important capability that must be augmented. This would be done through the creation of a series of security force recruiting centers located in key regions. These centers would be multinational facilities specifically designed to undertake the rapid training, education, and professional development of the security forces of a post-conflict state, including the military, gendarmerie or civil guard, police, intelligence services, national security civil servants, and judiciary.

Four Cultural Syndromes

Researchers identify four cultural syndromes, namely complexity, individualism, collectivism, and tightness. All four are patterns of beliefs, attitudes, self-definitions, norms, and values organized around some theme that can be found in every society. Using cultural syndromes as a frame of reference, we can develop foundational normative values having common application across all cultures, which should provide the starting point for a cultural education program. Cultural norms are often so strongly ingrained in daily life that individuals might be unaware of certain behaviors. Until they see such behaviors, they might have difficulty recognizing and changing them. Foundational cultural norms are normative values and factors having the greatest effect on military operations and the relations of soldiers with the populations they encounter.

Foundational Patterns:

- Communication styles
- Attitude toward conflict
- Approaches to completing tasks
- Attitude toward personal disclosure
- Approaches to knowing

Usually, our own culture is invisible until it comes into contact with another culture. People are generally ethnocentric: they interpret other cultures within the framework of the understanding they have of their own. This means studying differences in the domain of:

- Communication and language
- Dress and appearance
- Food and eating habits
- Time consciousness
- Relationships, Values and norms
- Beliefs and attitudes
- Mental process and learning styles
- Work habits and practices

I will only give you a short outline of communication and attitudes toward conflict: Communicating between two cultures involves generating, transmitting, receiving, and decrypting coded messages or bits of information; it is about much more than language, although language is certainly key to communication and should be a part of any cultural training program. The early focus, however, should be more on effective use and application of language than on making a soldier a linguist. Someone struggling to communicate in an unfamiliar language cannot communicate complex issues. The goal should be to orient language-skill developmental programs, at least initially, on effectively conveying simple terms (rather than on an acquisition of greater linguistic competence) so as to make the most out of simple meanings. The Army needs to find simple ways of communication that will speak to other cultural norms and that will require listening. Moreover other types of languages include mathematics, music, computing, physics, and engineering. Although such are not immediately useful in most military tasks, they offer a common frame of reference of possible value under special circumstances. One of the most overlooked and effective communication tools is using pictures, drawings, or photographs. A great deal of truth is behind the expression "a picture is worth a thousand words." Creating graphic and visual aides for cross-cultural communication is much easier (and often much more effective) than creating linguistic aides.

However, in any form of information transmission, meanings are not always clear, and certainly, missing presentation skills, timing, and context can be very confusing and counterproductive.

Attitudes Toward Conflict

Some cultures view conflict as a positive thing; others view it as something to be avoided. In the Western Communities conflict is not usually desirable, but people most often deal directly with conflicts as they arise. For example, a face-to-face meeting is a customary way to work through problems. In many Eastern countries, open conflict is considered embarrassing or demeaning. Differences are best worked out quietly. A written exchange might be the favored means to address the conflict. Another means might be enlisting a respected third party who can facilitate communication without risking loss of face or being humiliated. The military culture in general deals with problems head on. Many other cultures, however, employ indirect approaches and subtle means as part of day-to-day activity. When soldiers trained in the direct approach encounter these cultures, communication is difficult and can often lead to profound misunderstandings and miscalculations.

Cultural/Situational Awareness

Puts a premium on listening and comprehending the intent behind other's remarks under the aspect that our culture is invisible until it comes into contact with others.

Everyone is embedded in a culture that constitutes the manner in which one views others, the world, and oneself. Like an iceberg, some aspects of culture are visible; others are beneath the surface. Invisible aspects influence and cause visible ones.

Ethnography, a qualitative research method anthropologists use to describe a culture, attempts to fully describe a cultural group's various aspects and norms in an attempt to understand the group as such. The intent behind military cultural education is to help soldiers be more effective in the environments in which they must function but in which they have not been educated since their childhood. Therefore they must be culturally literate and develop cultural expertise for specific areas and regions. When balanced with study in potential areas of application, a solid cultural awareness supports the understanding of cultural factors in the areas of future operations. Usually, our own culture is invisible until it comes into contact with another culture. People are generally ethnocentric: they interpret other cultures within the framework of the understanding they have of their own. Fundamental patterns of cultural norms have greatly affected encounters between differing cultures in the domains of communication styles, attitudes toward conflict, approaches to completing tasks, decision-making styles, attitudes toward personal disclosure, and approaches to knowledge.[2]

Superiority over our Opponents[3]

In order to be superior to our opponents we must improve our learning climate in general. This implies being superior in the art of learning and adaptation.

Especially in: Communication Styles / Attitudes toward conflict / Approaches to completing tasks / Decision-making styles / Attitudes toward personal disclosure / Approaches to knowing (knowledge)

And what about ethics? The overall issue is: "Ethics and education of soldiers of all levels and ranks". But the more important question seems to be how to apply ethics and culture in the field?

I would like to investigate this issue by help of a „key model", FM 22-100, Army Leadership. The question is: Are there possibilities to learn about ethics/ culture in the field? I am quoting Robinson, Ethics Training and development in the military, "The US Military Academy takes on great efforts to instill essentially Aristotelian virtues into the cadets in order to build a character worthy of trust." And I would also like to quote Prof. Asa Kasher. Since the focus of most West-

2 Jean Willis, "Understanding Cultural Differences," American Society of Association-Executives, Washington, D.C., 1 March 2001.

3 Adapt or die, US Army and Training and Doctrine Command; White paper; Draft as of 06/10/2005

ern Armed forces (as well as the justification of most recent wars) has been humanitarian intervention and peace keeping operations, Kasher (qtd. in Robinson), postulates an appropriate combination of military and police ethics. The latter, he maintains, requires a different set of virtues. Maybe this is an appropriate solution. But as I tried to explain before, these operations do also require different skills and competencies.

Army Values and Leadership[4]

Army values in the traditional sense can also be an impediment to the transition of forces. We must launch a considerable amount of discussion as well as an increasing awareness of the need to transition and to a different self-understanding and, indeed, a different mix of unit-types and equipment than that for the large combined-arms fight, to accommodate the changed strategic environment. To state it at its extreme, the military in general has a significant resistance to embracing OOTW missions.

According to FM 22-100, these Values are to be observed: Loyalty / Duty / Respect / Selfless Service / Integrity / Personal Courage

Hilliard Aronovitch, journal of applied Philosophy, 2001 argues that: "Effective fighters are also ethical fighters, good soldiers in the one sense are also good soldiers in the other sense... Hence, good soldiers must in certain ways be good persons as well, Cit. in *Good Soldiers, a Traditional Approach.*" And I would like to add "Only inter-culturally competent soldiers can be good soldiers in this double sense."

The decision, that peacekeeping or nation building operations may require continued presence of military forces for decades or even generations is not a military issue. But assuring the readiness for such operations lies in the responsibility of the military leaders. But the key phrase in the minds of many officers, the foundation of their professional self understanding, is that the Army exists to win wars and not to win the minds of the civil population in foreign countries with fairly unknown cultures.

Asa Kasher, author of the "Code of Ethics of the Israel Defense Forces", has brought another aspect into discussion. He maintains that leaders should not try to build up an appropriate character in their subordinates. Developing the necessary appropriate identity in their subordinates will be more important. The main goal is to create an awareness of what it means to be a soldier in a liberal democracy and to fight for those values and beliefs wherever and whenever it is necessary. Then

4 See Leadership, Section I, character: what a leader must be, FM 22-100

Asa Kasher further develops the idea that for preparing a force for humanitarian interventions and peace keeping and enforcement operations there is a distinct demand for the right mixture of military and civilian ethics. This creates a completely new dimension for the transition of forces. This means recurrent learning processes for all levels in the army and is mostly a mental process.

In the light of these dynamics in this presented "Activity System" we have to give up the imagination of a learning process of a continuum from novice to expert as a simple linear process of internalizing as much of the necessary knowledge. We do, on the contrary, need to distinguish between internalizing of ideas artifacts and procedures already developed and their externalization as a creative new act as the learning process is going on. Both belong to experience and practice as long as practice is understood as a meaningful collective activity, not only as individual rehearsing of learned skills.[5] These two processes of internalization and externalization can be seen as recurrent processes.

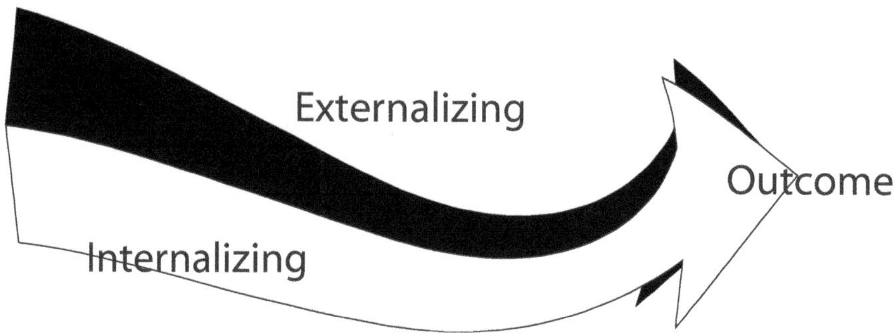

Figure 1. Ting- Toomey's mindful identity negotiation.

S. Weber had the idea to combine the "activity theory system" with the "mindful identity negotiation" framework.[6] In her opinion, it is precisely because she also places major emphasis on the aspects of identity, that Ting-Toomey's (1999) heuristic framework of "mindful identity negotiation" fits into the system.

5 Engeström, 1992,p. 15 cit. In S. Weber , Intercultural Learning as Identity Negotiation, Verlag Lang, 2005, J.v. Buer, Ed.

6 S. Weber, Intercultural learning as Identity negotiation, 2005, Verlag Lang

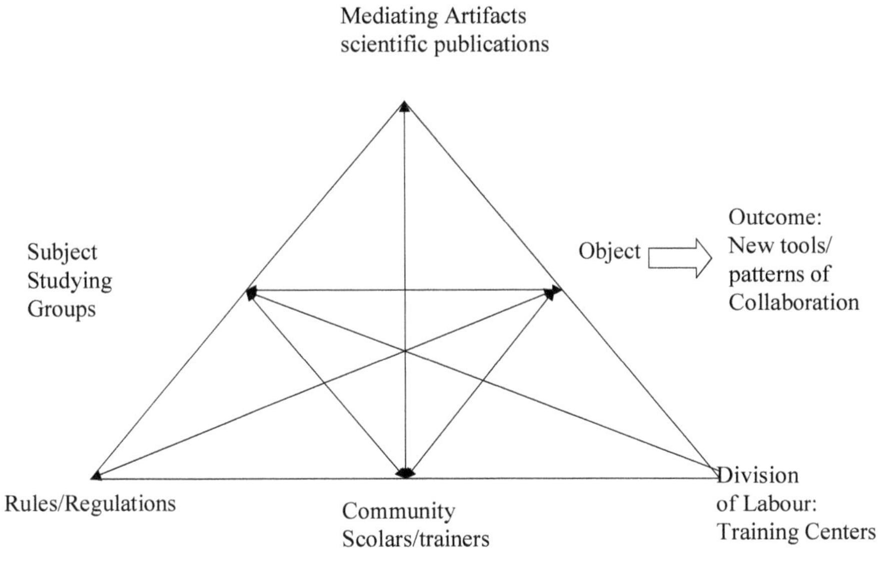

Figure 2. Intercultural framework.

Before explaining the system in greater detail, it is necessary to describe the notions of each basic element:

- Subject refers to the individual or group whose point of view is adopted in the analysis.
- The object refers to situation–bound goals to be achieved by certain actions, e.g. categories for describing intercultural interaction situations, necessary organizational plans directed to the activity.
- Instruments or tools refer to artifacts as human products e.g. cultural objects, signs or symbols, language, mental concepts, social institutions.
- Rules refer to explicit or implicit regulations that determine or constrain actions.
- Community refers to all participants in an activity system who share the same object/aim.
- Division of labor refers to the distribution of tasks, authority, or benefits of the activity system.
- The outcome refers to the transformed objects, such as new patterns of collaboration, new intellectual tools, concepts, the theories etc.

The central idea of activity theory is mediation of human action through cultural artifacts. It means that the subject and the object are mediated by instruments or cultural artifacts, such as symbols and representations of various kinds. But at the same time, the activity is also mediated by less visible social mediators like rules, community, division of labor.[7]

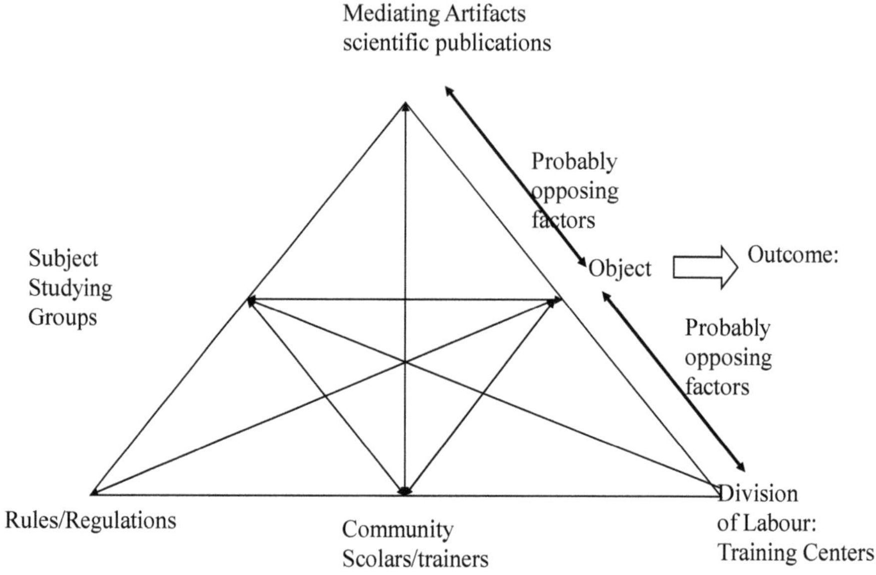

Figure 3. Tensions and contradictions (Engeström, 1999).

The tensions and contradictions, which arise between the elements of the activity system are the motivating force of change and development. Thus, activity systems are characterized by continuous transitions and transformations between the components. The activity system does not exist in a vacuum. It is rather one node in a multidimensional network-interdependent activity system.

7 Engeström, 1992, p. 12

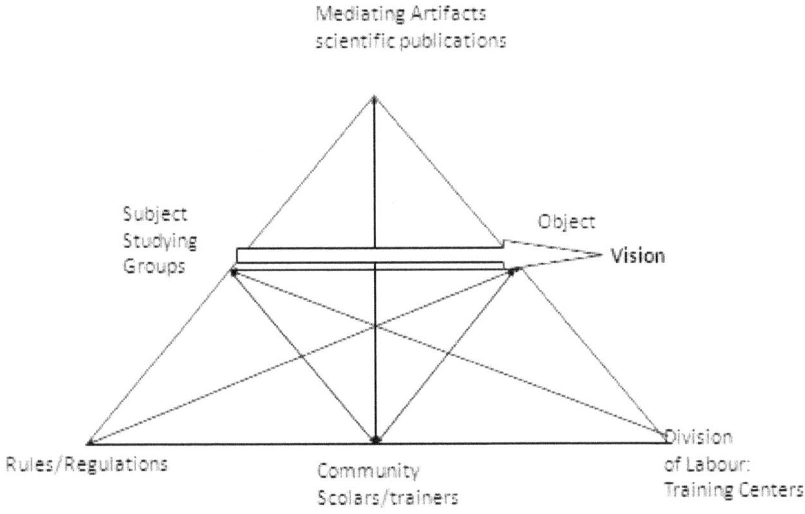

Figure 4. The aim – a common vision (Engeström, 1999).

Understanding Cultures[8]

Ting-Toomey draws upon work that discusses communication between Eastern and Western cultures, specifically between Asia-Pacific cultures and US and Canadian cultures in terms of individualism and collectivism. Cross-cultural psychologists, management people, and communications people have researched the dynamics of individualism vs. collectivism. The findings, below, are sources in Professor Ting-Toomey's research.

Understanding Cultures: Four Dimensions of Organizational Management Practices

Geert Hofstede[9] has been doing research in the international management area over the past twenty years. His latest database includes US-based multinational subsidiaries consisting of organizational management practices in fifty countries.

8 „Cross-Cultural Face-Negotiation: An Analytical Overview"by Professor Stella Ting-Toomey, presented on April 15, 1992
9 Hofstede, G.,1993, InterculturelleZusammenarbeit, Wiesbaden: Gabler

Hofstede discusses four dimensions in understanding organizational management practices:

1. individualism vs. collectivism,
2. power distance,
3. uncertainty avoidance
4. masculinity vs. femininity.

1. *Individualism vs. collectivism*. This dimension centers on organizational practices in individualistic cultures such as Canada, US, Australia, and Great Britain contrasted with collectivistic cultures in East Asia (Japan, South Korea, Hong Kong, and Singapore).
2. *Power distance*. Low power distance (Canada, US) subscribes to equal power distribution versus high power distance (Japan, South Korea, Hong Kong, etc.) in hierarchical structures.
3. *Uncertainty avoidance*. Hofstede found that Canada and the US are low in uncertainty avoidance, i.e., we like to take risks, take individual initiative, and enjoy conflict. Whereas cultures like Japan, Hong Kong, and South Korea are high in uncertainty avoidance, i.e., do not like conflict, but pursue group harmony; people within these organizations need clear rules, procedures, and clearly defined job responsibilities.
4. *Masculinity vs. femininity*. This dimension has been controversial because many people feel it is sexist. Hofstede discovered that Japan rated high on masculine dimensions (males expect an "in–charge" role). In contrast, countries like Norway and Sweden have a stronger feminine dimension, which means that roles are more fluid between males and females. Canada rated high on the masculine dimension compared with many Northern European organizational practices.

Individualism vs. Collectivism

Individualism vs. collectivism appears to be the most useful of the four dimensions. Professor Ting-Toomey (and her colleagues), Michael Bond, Harry Triandis, and Geert Hofstede consistently found that the individualistic and collectivistic dimension teaches the most about differences between cultures, particularly between East Asian and western cultures.

Individualism and collectivism is connected with the concept of identity, i.e., how do we conceive our sense of self? Individualistic cultures emphasize the "I" identity and collectivistic cultures emphasize the "we" identity, which is a fundamental difference between western and eastern cultures. The relational and communication patterns of this dimension are discussed below:

Relational Patterns

The communication process in individualistic cultures focuses on inter-individual levels, while collective cultures focus on the group base (whether you are in-group, one of us; or out-group, one of them). However, individualism and collectivism is not a polarized dimension: cultures function on a continuum. For instance, within the US or Canada, there are pockets of individualistic ethnic communities and pockets of collectivistic ethnic communities. So when discussing the individualistic versus collectivistic dimension, we are talking about patterns or value orientations within a country, in which there may be many variations. Pointing out the patterns and understanding them is a first step in learning to respect the differences.

Communication Patterns

In individualistic cultures, people tend to be verbally direct: we value communication openness, learn to self disclose, like to be clear, straightforward, and contribute to a positive management climate. In collectivistic group-oriented cultures, indirect communication is preferred because the image of group harmony is essential. In western cultures, talking is very therapeutic; in Asian cultures, there is an emphasis on observing and reflecting about the process. It is rare in Asian cultures to have open conflict, because it appears to disrupt group harmony.

Face Negotiation

Professor Ting-Toomey's work focuses on face-negotiation and conflict styles. She sees face as symbolic and as a claimed sense of self-respect in a relational situation. Face is a universal phenomenon because everyone would like to be respected; everyone needs a sense of self-respect. But how we manage the strategies in maintaining, saving, or honoring one's face differs across cultures.

Low versus High Contexts

The low versus high context communication framework was developed by Edward T. Hall.[10] Low versus high contexts (used similarly to individualism and collectivism) focus on the communication process itself. Low context centers on individualistic values and stresses verbal communication; while the high context system centers on group orientations and stresses non-verbal communication. Immigrant cultures with a pioneer spirit such as Canada, the US, and Australia tend to exhibit low context communication styles. Whereas high context cultures seem to have a longer continuity and history of culture; so for them, whenever they see a historical, cultural, or relational context, it is important to understand the communication process.

The low versus high context framework in face-negotiation has been used in Professor Ting-Toomey's research in the US, Japan, South Korea, China, and Taiwan. She has also collected data from Australia, Thailand, and India. Her research consists of open-ended questionnaires and interview data with nearly 1500 students and colleagues. Some of the concepts she explores are outlined below:

Face Saving

The question was asked, "What is the meaning of face?" American students tended to equate the concept of face with saving their own face, i.e., pride, reputation, credibility, and self-respect (relating to the ego). For them face is more individualistic, low context, and is associated with intra-psychic phenomena. Japanese students, on the other hand, understand the concept of face to be related to honor, claimed self-image, and the family/organization. For them there is more awareness of relational dynamics in the concept of face saving.

Face Giving

American students could not offer a meaning for face giving, whereas Asian business people/students could talk about the meaning of this term. Therefore, we may assume that face giving is more of an Eastern concern. To Asians, face giving

10 Hall, E.T.,1976, Beyond culture, New York: Doubleday, cited in S. Weber, Intercultural learning as Identity negotiation, 2005, Verlag Lang

means allowing room for the other person to recover his/her face-room to maneuver, to negotiate, so that both can gain face in the end. For Westerners, face seems to be a dichotomous concept: we either lose face or save face. For Easterners, face is considered to be a mutual, interdependent concept, and is a relational and group phenomenon. Public face is a concern in at least two-thirds of the world's cultures. But in Western cultures, there is also a sense of individual self-respect, the right to be treated with respect, and not to be embarrassed.

Face Losing

For Americans, loss of face means personal failure, loss of self-esteem, or loss of self-pride on an individual attribution basis. Whereas for Japanese and Korean subjects, loss of face means disrupting group harmony, bringing shame to their family, classmates, or company.

Recovery from Face Loss

For Americans and Canadians, humor is a strategy used to recover from face loss; if that does not work in a serious situation, other strategies that may be used are defensive strategies and attack strategies-clear win-lose strategies. Asian cultures, on the other hand, focus more on maintaining the image of a win-win process.

Conflict

Conflict is face-related, in that, face appears to be a predictor of what conflict strategies are being used. American subjects, for example, tend to adopt self-face preservation and maintenance, focus on self-face issues, use control-focused conflict strategies and confrontational strategies, and display stronger win-lose orientations. Asian subjects, on the other hand, tend to use face-smoothing strategies, mutual-face preservation strategies, and conflict avoidance strategies. People from Taiwan, Hong Kong, and the People's Republic of China use more compromising strategies; Japanese use more controlling styles (i.e. competitive) with out-group members and conflict avoidance with in-group members.

Face Issues

The word *shame* occurs when feelings are associated with face issues: shame in relation to self, parents, or group situations. However, practices differ between Japanese and Chinese: for the Japanese, if they disgrace their organization, they may also disgrace themselves and their family; for the Chinese, any disgrace reflects on their family honor.

Dimensions of Face

American subjects center on autonomy face (freedom) where privacy or personal space is important; high context, collectivistic cultures focus rather on "approval" face. In low context cultures, individual (intrinsic) accountability is emphasized; in high context cultures, however, group accountability or group dynamics may dominate.

Theoretical Implications

Ting-Toomey has provided a construct for understanding Asian business practices and communication style differences in order to develop other constructs for theoretical and research ideas. Some ideas are conceptualized below:

1. More multi–cultural team research efforts are needed. We should pay more attention to indigenous cultures, and look at their attitudes toward business practices and communication style differences. We need to identify core constructs with which we are unfamiliar, and are in danger of overlooking. Increased sensitivity is important because of global economic changes.
2. In training people to go overseas, it is critical to teach about low versus high context cultures as part of their applied skills. Listening skills are important in high context cultures. (In Chinese, the character for listening implies listening with your ears, eyes, and heart.) When collectivistic people come to individualistic cultures, they should learn to share more information, to self–disclose, to seek information more openly, and to seek to reduce uncertainty more than they normally would.
3. Within the face–negotiation framework, identity is important: How do we see the self? Who constitutes the in–group and the out–group? How do we move from out–group to in–group circles? What do we need to do in moving from ritualistic to substantive behavioral acts?

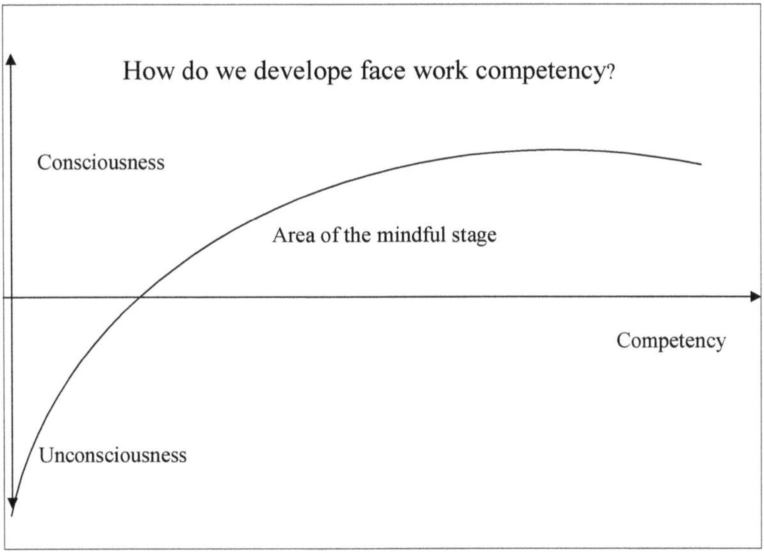

Figure 5. Face-work competency.

4. How do we develop face–work competency? There are four stages in communication competency:
   a) unconscious incompetence implies fundamental ignorance on the cognitive and behavioral levels;
   b) conscious incompetence means that behavior is understood but not synchronized, e.g., aware that there are too many awkward pauses and silences;
   c) conscious competence or the "mindful stage" means cognitively understanding communication differences, the patterns and variations of different cultures, and working on behavioral face work competence.
   d) unconscious competence is rather like driving a car or swimming; at a certain point, it becomes spontaneous, natural. Spontaneity is part of practicing face work, you adjust and you adapt. You become "mindlessly mindful" at this final stage

Motives

- Socio–economic conditions, multicultural attitudes, degree of institutional support, in and out–group definitions, degree of cultural differences.
- motivational expectations, cultural knowledge
- existence of networks, ethnic media, interpersonal adaptation ability[11]
- cultural knowledge
- knowledge about one's own culture as well as others' is the most critical component in managing intercultural differences.
- communication skills

Ting-Toomey suggests various interaction and communication skills:

- "mindful" observation
- "mindful" listening
- practicing verbal empathy
- practicing identity confirmation
- face work management skills
- building up trust
- collaborative dialogue skills
- self reflection

Mindful identity negotiation

Mindful identity negotiation requires a high degree of consciousness and reflectivity about the elements and processes of an intercultural encounter. This includes becoming aware of one's own history, value orientations, views, identity needs, etc., as well as questioning one's own experienced behavior and practices.[12]

This process seems to be crucial to intercultural relationship and in any form of intercultural learning. The confrontation between different patterns if interpretation and practice leads the individuals to some kind of self-reflection so that they recognize that there are alternative ways to handle a new cultural phenomenon. It is also necessary for the individuals to understand their own identities.

11 Ting-Toomey, 1999, p. 233-260, cited in S. Weber, Intercultural Learning as Identity Negotiation, Verlag Lang, 2005

12 Ting-Toomey, 1999, pp. 8, 40, cited in S. Weber, Intercultural Learning as Identity Negotiation, Verlag Lang, 2005

The discussion, in which both partners or groups present, evoke, challenge, and, or support their own and other's desired identities, is often characterized by disharmony. It helps the individuals to come to the insight that there be something to be done. This corresponds to the first phase of the expansive learning cycle. To gain more insight into the alien patterns of interpretation and thinking, the individuals develop adequate instruments for mindful observation/listening. This information enables them not only to figure out the identity needs, attitudes, cultural orientations and further cultural knowledge. They then integrate and transform their extended cultural knowledge with other pending issues developing relationships and/or conflict styles to develop mindful intercultural skills for the ongoing collective activity. When the partners achieve a commonly shared meaning through negotiation and reach their goals, which means, they feel understood, respected and supported, they will judge their negotiation process as appropriate, effective and satisfactory. So the partners develop a so-called inter-culture.[13] This phase corresponds to the third phase of expansive learning cycle creating a new vision. The whole process has to be repeated in each new encounter but for economic reasons the partner groups routinize this process. But routines must be controlled from time to time to avoid misunderstandings and failures within the activity system. Furthermore, necessary adaptations need to be implemented in to the model every now and then.

13 See Bolton,1995,Grenzen der Internationalisierungsfähigkeit. Interkulturelles Handeln aus interaktionstheoretischer Perspektive, cited in S. Weber, Intercultural Learning as Identity Negotiation, Verlag Lang, 2005

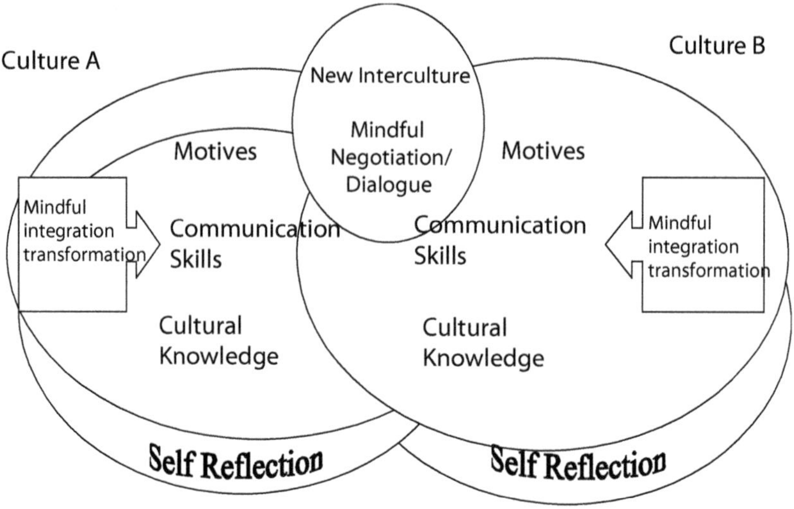

Source, slightly altered: S. Weber, Intercltural Learning as Identity Negotiation, Verlag Lang, 2005

Figure 6. Emerging a new interculture (based on Weber, 2005).

References

Beyerchen, A. (1997). Clausewitz, Nonlinearity, and the Importance of Imagery. In D. Alberts & T. Czerwinski (Eds.), *Complexity, Global Politics, and National Security* (Chapter 7). Washington, D.C.: National Defense University.

Fastabend, D. & Simpson, R. (2004). Adapt or Die. *ARMY Magazine*, 52 (2), 15-21.

Hooker, R.D. (2005). Beyond Vom Kriege: The Character and the Conduct of Modern War. *Parameters*, 35 (2), 4-17.

Metz, S. & Millen, R. (2005). Intervention, Stabilization and Transformation Operations: The Role of Landpower in the New Strategic Environement. *Parameters*, 35 (1), 41-52.

McFarland, M. (2005). Military Cultural Education. Army Review, March-April, 62-69 [Online]. Available: http://www.army.mil/professionalwriting/volumes/volume3/june_2005/6_05_3.html (Retrieved on June 2009).

Sargent, R. (2005). Strategic Scouts for Strategic Corporals. *Military Review,* March-April, 12-17.

Wallace, W.S. (2005). Network-enabled Battle Command. *Military Review,* May-June, 2-5.
Weber, S. & van Buer, J. (2005). *Intercultural Learning as Identity Negotiation.* Frankfurt: Lang.

2. Training and Education from National Perspectives

Hubert Annen

Current challenges of Swiss military pedagogy

Some mental gymnastics

If one is to look back at the military pedagogy conferences that have taken place so far (see Toiskallio, 2000; Micewski & Annen, 2005), several central contents can be identified. Basic questions of *training*, *leadership* and *education* are discussed, as well as the pedagogical challenges that emerge as a consequence of technological progress or political developments. In terms of the latter, the concrete concern is primarily with the growing importance of military ethics, the behaviour in a multicultural environment as well as the consequences of network centric warfare on leadership and communication. The situation can therefore be compared to that in the civilian realm, where educational science is equally experiencing a practically *seismographic* reaction to the constantly changing societal problems (Gudjons, 1999).

It might then be argued that while this is all right and proper, military pedagogy nevertheless still only functions as a sideline. At first glance, this might indeed be the case, for if one follows the coverage and analyses on army deployments from around the world, political, strategic or technological aspects are in the foreground. However, upon closer inspection, it can be established, for instance, that the dream of "effects-based and net-centric operations, where sensors, computers and telecommunications networks would lift the fog of war" has arguably gone up in the same smoke. Actions of the enemy in Iraq have made the techno warriors about as credible today as stockbrokers after the Great Depression (Scales, 2006). Political or strategic decisions are also frequently based on foundations other than those that ultimately determine a possible success in the crisis region concerned.

A possible answer to the question of where success factors can nowadays be found is provided by the historian Alan Beyerchen (quoted from Scales, 2006). He has developed a taxonomy of war in the modern era in terms of four world wars. Each war was shaped by what he calls "amplifying factors". For example, World War I was a *chemists'* war in that the decisive strategic advantage on the battlefield was driven in large measure by new applications of chemistry and chemical engineering. The war should have ended for the Germans in 1915 when their supplies of gunpowder nitrates exhausted. But the synthesis of nitrates by German scientists allowed the war to continue for another three horrific years. World War II was a *physicists'* war. To paraphrase Churchill, the atomic bomb ended the conflict, but exploitation of the electromagnetic spectrum in the form of the wireless and radar won it for the allies. The Cold War – also referred to as World War III – was the *information researchers'* war, a war in which intelligence and knowledge of the enemy and the ability to fully exploit that knowledge allowed the U.S. to defeat the Soviet Union with relatively small loss of life. And finally World War IV- the time after 9/11 – is the *social scientists'* war. To win this war, the military must be culturally knowledgeable enough to thrive in an alien environment. Victory will be defined more in terms of capturing the psycho-cultural rather than the geographical high ground. Understanding and empathy will be important weapons of war (Scales, 2006).

If this is the case – and there is a great deal to indicate that it is – it consequently becomes clear that *human factors* have gained decisively in importance. Cultural awareness, tactical intelligence, psychological knowledge, leadership, and intuition are abilities that have always been important, but which for the first time should tip the scales towards victory or defeat. With the increasing importance of *social skills*, the social sciences are given more weight in the military context. Against this background, it can then also be postulated that military pedagogy should in no way be seen as being on the sidelines.[1]

On the whole, Beyerchen and Scales provide grist to the mill of all employees in military organisations who work in the area of the social sciences. However, military superiors do not like to be interfered with when it comes to their core business of leading their subordinates – and certainly not by some academic "pencil pusher" or other. This is again confirmed by Scales (2006): "The relationship between the military and human and behavioral scientists has, to date, been one of antipathy and neglect. Academics and behavioral practitioners have rarely violated the turf of the soldier." We therefore see ourselves in general terms as being faced with the challenge of how we can implement a social scientific body of thought, i.e. in our concrete case military pedagogical thinking, in such a way that it is accepted by the users and is conducive to military task fulfilment and goal achievement.

Approach on two levels

If one looks at military pedagogy from a *strategic* perspective, the concern is above all with the *contents* spawned by the current threat situation and the demands on

1 As military pedagogy cannot be clearly delimitated from other social scientific subjects taught in the military, in this contribution, military pedagogy is always also meant when social sciences are referred to.

military organisations derived from this. It is also a question of psychological strain, whether and to what extent it suddenly becomes vital to draw on the findings of social scientific research and practice.

When, in the 1st World War, the fitness for service of several hundreds of thousands of American soldiers had to be tested within the shortest space of time, a link to the psychologists was quickly produced – and with the specially developed psychometric tests it was indeed then possible to make the necessary screening decisions efficiently and more or less reliably (Geuter, 1999). As a further example, it can be argued that armies have seen themselves as increasingly confronted with psychological casualties (combat stress reaction / posttraumatic stress disorder) since the beginning of the 20th century. Here too, it was impossible to avoid drawing on the know-how of psychologists in order to cope with this fact (see Bartone, 1998). In both cases, incidentally, the relationship proved to be profitable for both sides: The development of the Army Alpha test in around 1915/1916 represented the first major acid test for modern psychology, and at the time, the army was one of the most important employers for psychologists; and with regard to the question of stress management, military organisations provide arguably one of the most interesting fields of research and occupation for psychologists. Another current example that can be cited is the reaction of the US Armed Forces to the accidental shooting down of an Iranian passenger aircraft, where a group of researchers was immediately commissioned to examine the fundamental factors of decision-making under stress (Cannon-Bowers & Salas, 1998). And finally, the interest in getting a better understanding of a soldier or of an officer as an information-processing system can be understood as a reaction to Network Centric Warfare.

As mentioned at the outset, with regard to military pedagogy, several comparable trends can also be discerned. Nowadays, one sees oneself on the whole as being confronted with asymmetrical warfare or operations other than war, i.e. the threat mostly presents itself in a very diffuse manner and the individual soldier often finds himself working in and having to prove himself in grey areas. In addition, he has become much more important as an autonomous part of a whole system. As the example of Abu Ghraib or other publicly noted maltreatments by army members against civilians show, certain acts carried out on the group level can take on strategic dimensions. Against this background, it is not surprising that on the one hand questions of military ethics and the resulting transmission of values, intercultural competence and interpersonal communication are placed in the centre of military pedagogical thought and action, and on the other hand, holders of responsibility in the military are again taking an increased interest in military pedagogy.

All the same, it can be shown by the curriculum for the career officer training in the Swiss Army that this fact does not only correspond to the wishful thinking of a military pedagogue (Annen, 2003; Annen, Steiger & Zwygart, 2004). Ten years ago, the three-year diploma study course for career officers received two semester periods per week on leadership in the first year, and also two semester periods per week of military sociology in the third year. Today, social scientific subjects such as military pedagogy, psychology, military psychology, leadership and communication management take up approximately one third of the course plan in the first year of the training, which has been graded up to a Bachelor's degree. In addition, in the further course of the degree, contents are consolidated in several weeklong block courses in a targeted and practice-oriented manner. In this regard, it should be emphasised that military pedagogy is seen as a core subject and is weighted accordingly in the examinations.

This quantitatively discernible upgrading of the social sciences – and therefore also of military pedagogy – is without doubt gratifying. However, it would be negligent to use this fact to rest on one's laurels. In the arrangement of the syllabi, it can at times be observed that those responsible run the risk of succumbing to a certain "checklist mentality". They grant the social scientists the required scope in order to show to the outside world or the powers that be (i.e. political supervisory bodies) that "something is being done". The duty has been done and everybody can relax. In reality, this is where the real work begins, for it is necessary to fill the vessels concerned with content and to ensure the transfer of theory to everyday practice. The fact that this is not self-evident is shown by current and comprehensive investigations, which have demonstrated that the effect of study courses in the area of "social skills" tends to lie at around zero in most cases (see Frieling et al., 2000).

It is therefore vital to contemplate the *operational* level of military pedagogy. Here, the concern is with everyday business and consequently with concrete *methods*. Based on solid basic knowledge, it needs to be thoroughly considered how the relevant contents can be imparted in order for them to find their way from the theory room into practice. In this regard, there are clearly fewer laurels to be gained than in political discussions or at international conferences, where the state of affairs does not necessarily have to be proven. Military pedagogical work at grass-roots level requires commitment and presence, and as is common in terms of educational work, both successes and failures are directly palpable.

Using concrete examples, in the following, the basic military pedagogical work will be illustrated in the framework of the cadre training in the Swiss Army. At the same time, it will be shown how the modules concerned are evaluated in order to ensure as great a sustainability of the inputs as possible.

Military pedagogy in the cadre training of the Swiss Army

The Swiss Army continues to function as a conscript army. Following a two- to three-day physical and psychological fitness screening, just slightly fewer than 60% of the 20-year-old men are drafted into military service. The remaining men carry out their compulsory service in civil protection or in alternative national service, or if deemed unfit for service, have to pay a substitute tax. In the first seven weeks of the basic military training centre, approximately one third of all recruits are selected for a cadre career track. Following a subsequent ten-week cadre school, it is decided who will be accepted for officer training. The prospective officers then complete a five-week officer candidate school, before they are enlisted in the four-week central officer-training course, where all officer candidates of the Swiss Army are brought together. Here, above all basic contents in the area of Swiss security policy, leadership and military values are imparted. Following this, it is necessary to pass a 20-week officer school in the candidate's own training unit (e.g. infantry). As the final element, this is followed by a practical service lasting for eight weeks, where the officer, who has been freshly breveted to lieutenant, leads a platoon independently. Well-qualified officers who have a Matura (Swiss university entry-level qualification) have the opportunity to pursue the track of career officer. The 50-100 candidates per year have to undergo further selection procedures (sport, social skills), and approximately 30-50 of these then undertake the three-year Bachelor's degree at the Military Academy. In the Swiss Militia Army, the proportion of professional soldiers among the cadres is just under 5%.

Self-reflection

In the cadre school, a great deal of weight is already placed on systematic leadership training. The so-called "leadership training for lower militia cadres" comprises modules such as self-awareness, communication, conflict management, leadership psychology or personnel management. A certain number of successfully completed training modules qualify the cadre to achieve a certificate as a "leadership specialist" that is recognised in the civilian domain. The first module, "*self-awareness*", is congruent with the *appraisal system* according to which the prospective cadres receive regular feedback about their performance and behaviour. The instrument concerned was developed according to the principles of action research[2], i.e. under

2 Translation work between theory and practice is a central feature of action research. Accordingly, this method should be one of the most promising approaches in military pedagogical research work.

direct and constant inclusion of practitioners as well as considering academic findings (Annen, 2000). In a stepwise process, this enabled the creation of an appraisal procedure that contains the most essential behavioural characteristics of an officer of the Swiss Army, whereby once again, the self-competence and social competence is given a great deal of weight. In addition, all cadres and prospective cadres have to carry out a self-assessment on the same basis. The training module "self-awareness" is therefore given a practical and consistent continuation. Moreover, through the fact that the subordinate arrives at his appraisal interview with the completed self-assessment, this feedback session is given further meaning. Instead of being carried out in the form of "one-sided transmission" by the superior, a dialogue is carried out, which ultimately enables the development together of measures according to which the appraisee can improve.

The appraisal procedure concerned is continuously evaluated. After several teething problems, where military superiors struggled with the cultural change, and the extra effort involved with the differentiated form was criticised, acceptance is now quite good (Annen, 2005c). A comprehensive analysis of the completed forms as well as the feedback interview has firstly brought to light that the appraisees tend towards a certain overestimation of themselves, with this effect being greater in those evaluation dimensions relating to personality characteristics (Annen & Kamer, 2006). Secondly, it was possible to show that the effectiveness of the feedback interview, i.e. the motivation of the subordinate to work on the improvements discussed in a targeted fashion, strongly depends on the perceived credibility of the feedback giver.

With regard to the military pedagogical work, in general, the following can be established:

- Military pedagogical interventions are only as good as their acceptance and implementation by the users. In addition to the academic quality, the practicability therefore always has to be considered.
- Leadership is often perceived differently than how it was meant. The perspective on matters therefore has to be coordinated in a constant dialogue.
- Even the best appraisal instrument is only an aid. The success of pedagogical work, as here in the concrete example of the appraisal of subordinates, ultimately always depends on the people involved.

Clarification of values

As reported in earlier studies (Annen, 2004; 2005a), in the last few years incidences have repeatedly occurred in the Swiss Army that were carried to the general public, with particularly great gusto by the tabloid press. Mostly, the concern was with leadership errors, misguided group dynamics or questionable rituals in military schools and courses. As a consequence, at the beginning of 2005, the Commandant of the Armed Forces College commissioned a teaching module to be set up on the theme of "leadership with values". To this aim, in a first step, together with experienced leaders of different hierarchical levels, concrete cases were gathered that had occurred in the last few years. These were subjected to a qualitative content analysis by means of which the following central problem fields were analysed: "creative" educational measures, inappropriate training methods, demonstrations of power, using the wrong words and tone, group momentum and group dynamics, respectively, rituals and unassertive superiors. On this basis, the available material was compressed into eight model cases and developed from a didactical perspective, i.e. it was given a uniform structure and reference was made to relevant principles (military service regulations, military penal law, leadership guidelines). The product, which was made available to the trainers in the central officer-training course, additionally contained a methods manual with concrete suggestions for implementation in the teaching. The aim was to sensitise prospective officers to delicate topics by working on these cases. With this mental confrontation with the subject matter as well as the discussion of possible solutions, the danger is guarded against that in practice, they will be surprised and overstretched by a comparable situation; consequently, they should be in a position to react competently and appropriately.

The experiences with this teaching sequence are discussed with the whole teaching body of the officer-training course on a yearly basis, enabling possible optimisations to be derived. In concrete terms, it meant, for instance, that based on current events in terms of "Diversity Management", it was possible to implement a new problem field without delay. In addition, a study is currently in progress that is examining the short-term and long-term effect of the module "leadership with values". Initial results show that the cadets are able to position themselves more clearly in terms of their subjective opinion following the teaching sequence concerned. On the part of the participants, however, it is criticised that the time allocated for the topic is too short, and more interaction is also desired. In addition, it can be established that opinions on this module strongly depend on how it is implemented by the trainer in charge.

Accordingly, the following central military pedagogical findings can be established:

- A teaching sequence is never a finished product. It needs to be constantly reflected upon, enabling a timely reaction to current problems.

- The effect of military pedagogical training is difficult to discern. There is consequently the danger that the trainers will favour (more trivial) contents that are easier to measure and promise short-term training success.
- Once again, it is clear that the personality of the person imparting the information is of decisive importance for the effect of military pedagogical work.

Leadership culture

The assessment center for prospective career officers (ACABO) is the most essential selection hurdle on the path from militia officer to career officer. Here too, the emphasis is on the evaluation of self-competence and social competence. Therefore, in addition to having a functional aspect (personnel selection), this procedure also has an implicit aspect, through the fact that the targeted selection of credible, human-oriented leaders also shapes the leadership culture as such. Bearing in mind the importance of this selection instrument, the ACABO is regularly subjected to academic examination. Accordingly, we will not address this further at this point, but refer rather to the numerous publications on the matter (see Annen, 2005b).

Sustainability

As mentioned, in the framework of the three-year Bachelor's degree for becoming a career officer, a fair amount of weight is given to the social sciences in general, and consequently also to military pedagogy. However, a problem arises here that can arguably be observed in any "classical" degree course: The students only take in what appears relevant for the examinations; contents are learned mechanically, processed specifically for the examination and then quickly forgotten again. In short: testing drives instruction. Against this background, it is therefore not particularly surprising that in spite of the fact that the human factors are granted a great deal of training time, in practice, the same leadership errors are constantly observed. Military pedagogical contents should accordingly be examined in such a way that it is not the contents but rather the processes that are effective in everyday life that are in the foreground. In other words, if (self-)reflection, interaction and discourse are the central emphases in the teaching, then the performance appraisal should be congruent with this – i.e. *instruction drives testing*. For this reason, in the subject of military psychology and military pedagogy, the subject matter has recently started to be tested according to the portfolio method (see Jones, 1994; Bohl, 2005). In concrete terms, this means that the students already document their learning process during the semester on a continuous basis. They are compelled to assume responsibility for their own learning, recognise coherence and interconnections, and through this critically deal with the subject matter in such a way that the practical usefulness of the knowledge is brought to the fore. Accordingly, each student works on his individual portfolio, in which he records learning within and outside of the lessons, stores formal and informal documents and shows his standard of performance. Generally speaking, he tells his own learning story. In the portfolio interview at the end of the semester, it is then discussed what has been entered into the portfolio, why it was selected, how one rates one's own performance and learning progress and what practical consequences one thinks should be drawn. In addition, the lecturer poses several knowledge questions.

The experiences with this appraisal instrument are systematically evaluated and analysed. Currently, corresponding data from two study courses are available. Predominantly positively, it is noted in this regard that in comparison to traditional evaluation procedures, there is a greater reflection on learning behaviour, one's own learning process is perceived more consciously, and there is therefore also a greater awareness of one's own state of knowledge. The feedback on the portfolio interview turned out to be particularly gratifying. The impression given is that one is able to represent the state of knowledge well and that accordingly, the grade is clearly easier to comprehend than is the case, for instance, in a written examination. Ultimately, 75% of those surveyed were of the opinion that the portfolio enabled better conditions for a transfer into practice. Points of criticism were the additional effort required during the semester as well as the less than clear guidelines on what exact form the portfolio should take. As a consequence, although they are in line with the intention of the procedure, these two particular problems shall be more explicitly communicated to the students in advance in the future. From the perspective of the lecturers carrying out the portfolio process, it can be added that the phase of portfolio discussions is very intensive and requires a markedly high mental and emotional presence. Only in this way can justice really be done to the students and their individual learning history.

The method of portfolio assessment and the first evaluation results lead to the following generally relevant insights for military pedagogy:

- The teaching of interpersonal processes, as it is represented to a broad extent by military pedagogy, can be depicted with a content-oriented test only to an unsatisfactory degree.
- Military pedagogical teaching demands from the trainer an above-average mental and emotional presence; consequently, this is also the case when the students are being examined in this subject.
- Evaluation methods, be the focus on academic testing or a performance examination in a degree course, have to adapt to the subject. For the former,

the action research method has proved to be valuable, while for the latter the portfolio assessment is more expedient.

Link theory – practice

After a career officer has worked in practice in the Swiss Army for 5-8 years, with regard to his further career course, he has to complete an additional training course at the military academy lasting for 12 weeks. Among other things, this course offers the opportunity to refresh, reflect upon and discuss military pedagogical contents with these participants against the background of their practical experiences, which in turn can have concrete consequences for the military pedagogical teaching in the basic training. To this aim, as part of this year's course, the career officers in question were given the task of writing an essay with the title "The career officer as a military pedagogue". These essays were evaluated by means of content analysis and the relevant factors from the point of view of the experienced career officers can currently be presented as follows:

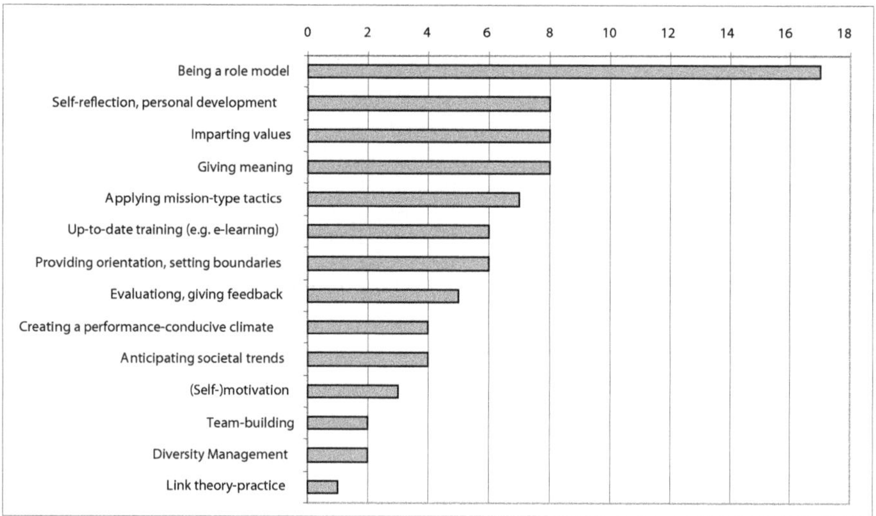

Table 1. The career officer as a military pedagogue – Key factors /challenges from the perspective of experienced career officers (N = 24; Maj/LT Col, 35 – 40 y.)

In view of this result, the general alignment in the basic training appears to be right. The further elements of military pedagogical teaching mentioned above are also found in the essays. As a direct consequence, e-learning as a subject and its meaningful implementation alone will be given more attention in the future.

With regard to military pedagogy, in general terms, the following emerges:

- More than ever, the (military) superior has to set himself the task of bestowing meaning and orientation.
- Military pedagogical work means above all creating a framework for making the subordinates able or willing to fulfil their task (mission-type tactics, work climate, team).
- The contents of military pedagogical accomplishments are largely given and experience only minor changes over the years. What is decisive, by contrast, is to create as concrete a picture as possible in the training regarding what it means in practice, e.g. to be a role model or to impart values, for the mere issuing of orders surely does not achieve this.

Conclusion

It is important to define the *strategic* level of military pedagogy. Through this, from the *meta-perspective*, trends and developments are identified that draw the framework for military pedagogical work. Efforts at persuasion are based on the same foundation, so that this theme, which is becoming increasingly important but which is difficult to comprehend for military leaders, can be granted the necessary space in the training.

On a somewhat provocative note, however, it can be commented that it is comparatively easy to lecture from a distance on, for instance, military ethics regarding all sorts of splendid and socially desirable factors. In addition, the content-related orientation of pedagogical functioning should stay roughly the same, as certain values such as consideration of human dignity, mutual respect or camaraderie are, for example, of general and temporally enduring validity. Things become more tangible, however, on the *operational* level, where, for example, the concrete question arises of how, against the background of current conflicts and army deployments, human dignity can or should be taken into account. Here, in everyday military pedagogical work, cases and dilemmas need to be discussed, and solutions need to be developed against the background of practical experiences and academic foundations. Often this requires personal commitment as well as the courage to go out on a limb. Military pedagogy cannot be organised or delegated, but needs to be understood as an enduring process, which has to be lived out and reflected upon. Its challenge is the translation work between theory and practice, and its benchmark is therefore the sustainability of teaching and consequently the *action competence* of individual army members in both training and deployment.

References

Annen, H. (2000). *Förderwirksame Beurteilung. Aktionsforschung in der Schweizer Armee.* Frauenfeld; Stuttgart; Wien: Huber.

Annen, H. (2003). „Nichts ist praktischer als eine gute Theorie" – Militärpsychologie und Militärpädagogik in der Schweizer Armee. In: Militärakademie an der ETH Zürich (Hrsg.), *Von der Militärschule zur Militärakademie. 125 Jahre Militärwissenschaften an der ETH Zürich* (S. 81-98). Au: Militärakademie an der ETH Zürich.

Annen, H. (2004). *Authority, power and obedience. An efficiency analysis of a teaching sequence.* Paper presented at the 5th International Military Pedagogy Conference, Swindon, UK, 13-15 December 2004.

Annen, H. (2005a). *Abuse of authority and misguided group dynamics – military pedagogical measures.* Paper presented at the 6th International Military Pedagogy Conference, SOWI Strausberg, Germany, 28 November – 1 December 2005.

Annen, H. (2005b). Geeignet als Berufsoffizier? – Das Assessment Center für angehende Berufsoffiziere (ACABO) in der Schweizer Armee. In: S. Collmer & G. Kümmel (Hg.) (2005). *Ein Job wie jeder andere? Zum Selbst- und Berufsverständnis von Soldaten.* Baden-Baden: Nomos.

Annen, H. (2005c). Mit Prinzipien und Offenheit zu einer glaubwürdigen Beurteilungspraxis. *Allgemeine Schweizerische Militärzeitschrift,* 12/05, 14-15.

Annen, H., Steiger, R. & Zwygart, U. (2004). *Gemeinsam zum Ziel. Anregungen für Führungskräfte einer modernen Armee.* Frauenfeld; Stuttgart; Wien: Huber.

Annen, H. & Kamer, B. (2006). Die Selbstbeurteilung im Qualifikationsprozess – erste Erkenntnisse. *Allgemeine Schweizerische Militärzeitschrift,* 2/06, 7-9.

Bartone, P. (1998). Stress in the Military Setting. In: Cronin, Ch. (Ed.), *Military Psychology – an Introduction.* Needham Heights: Simon & Schuster.

Bohl, T. (2005). *Prüfen und Bewerten im Offenen Unterricht* (3. überarb. Aufl.). Weinheim; Basel: Beltz.

Cannon-Bowers, J. A. & Salas, E. (Eds.) (1998). *Making Decisions under Stress.* Washington, DC: APA Press.

Frieling, E., Kauffeld, S., Grote, S. & Bernard, H. (2000). *Flexibilität und Kompetenz: Schaffen flexible Unternehmen kompetente und flexible Mitarbeiter?* Münster; New York: Waxmann.

Geuter, U. (1999). Militärpsychologie. In: H. E. Lück & R. Miller (Hrsg.), *Illustrierte Geschichte der Psychologie* (2. korr. Aufl.) (S. 279-283). Weinheim: Beltz.

Gudjons, H. (1999). *Pädagogisches Grundwissen* (6. erg. Aufl.). Bad Heilbrunn: Klinkhardt.

Jones, J. E. (1994). Portfolio Assessment as a strategy for self-direction in learning. In: Hiemstra, R. & Brockett, R.G. (1994), *Overcoming resistance to self-direction in adult learning* (p. 23-29). San Francisco: Jossey-Bass.

Micewski, E. & Annen, H. (Eds.) (2005). *Military Ethics in Professional Military Education – Revisited.* Frankfurt a. M.: Peter Lang.

Scales, R.H. (2006). Clausewitz and World War IV. *Armed Forces Journal,* July 06. [On-line] Available: http://www.armedforcesjournal.com/2006/07/1866019 (Retrieved on October 2006).

Toiskallio, J. (Ed.) (2000). *Mapping Military Pedagogy in Europe.* Helsinki: Finnish National Defence College, Department of Education.

Bo Talerud

Ethics, Ethos and National Culture

After the end of the Cold War, many of the western countries changed their security policy completely. From giving priority to defence of their own territory there has been a shift in focus to different kinds of international missions, mostly efforts to peace-reconstruction. To what degree have their military ethos and ethics adjusted to this new focus?

The Swedish peacekeeping force in Kosovo as a small illustrative case

As a start, I will refer to interviews with Swedish officers, privates and local interpreters in Kosovo in April and September 2007. According to those interviews and my own observations, the internal standard of morals at the camp is high in regard to alcohol, drugs, noise, equipment, care etc. The relations between men and women are relatively equal, even if there are still some traits of "macho ethics" among some of the younger soldiers. Most Swedish soldiers behave with openness, friendliness, fairness and respect toward the local population, but can also be firm if the situation becomes more demanding. Swedish soldiers in international missions are by selection relatively well educated, some even at a university level. They also volunteer for the missions.

As I understood it, KFOR in general and not least the Swedish peacekeepers are highly appreciated by the local civilians. But is this due to proper training and education in ethical matters within the armed forces?

Forming agents of Swedish society and national culture: The heritage from Swedish social movements and popular adult education.

Many of the abilities mentioned above, are more probably the results of the socialization into Swedish culture and society. Strong forming agents of Swedish society and national culture are the heritage from Swedish social movements, its popular adult education with its impact on the formal educational system.

During the 19th century a strong movement of popular adult education developed in Sweden and other Nordic countries. The achievements of the Danish historian, philosopher, clergyman and adult educator N.F.S. Grundtvig (1783-1872) were an important source of inspiration. Movements were also influenced by the foundations of the Danish Folk High School.

At the beginning of the 20th century, popular adult education connected with social movements, the Free Church, the temperance and the working-class movement developed. Within those movements the educational form of study circles became popular. These were informal group meetings around a topic or a book, with a self-chosen leader to run the discussion. Conscious self-directed education among the participants was important. Popular education also strongly contributed to the democratization of Sweden during the first half of the 20th century. The Government, the industry leaders, the labour unions and the social-democratic party all learned that by using dialogue instead of confrontation, all parts could contribute to the development of society, for the benefit of all (Bron 2004, Talerud 1985)[1]. So the Swedish model of the welfare state had its golden days from the 1930s up to the 1980s. The forms and methods of popular education have influenced teaching in the formal educational system,

I think that this heritage of self-directed education and internalization of democratic and humanistic values, at least indirectly, has had some impact on the Swedish Armed Forces.

The Swedish formal educational system

As early as in 1946, the official report of the National School Commission stated that the most important goal of the Swedish educational system was to "foster democratic human beings" (SOU 1948 p. 27)[2]. The central curriculum for comprehensive schools emphasized human dignity and that the UN's *Declaration of Human Rights* should be kept alive in discussions with the pupils.

The central curriculum for the comprehensive school emphasized these human aspects even more in 1980:

"Education shall aim at strengthening respect for basic human rights. Further, education shall promote preparedness for efforts to bring about peace and increase understanding of international solidarity. Education shall lead to understanding

---

1 Bron, Agnieszka & Talerud, Bo: Co-existence or Convergence of General and Vocational Education in Sweden. pp 35-38 in Lechner & Pöggeler (Eds): *Allgemeinbildung und Berufsbildung – Konkurrenz und Kongruenze der Konzepte im Europa des 20 Jahrhunderts*. Peter Lang Verlag 2004. Talerud, Bo: *Cultural Pedagogy in a Technological Society – On the possibilities of cultural development through pedagogical activities.* Doctoral thesis Department of Education, university of Stockholm 1985

2 1946 School commision report with recommendations for the development of the Swedish school system. Stockholm Ministry of Education

and respect for all people, for their cultures, civilizations, values and manner of living." (The Swedish National Board of Education 1980 p. 120, translated and quoted by Johansson 2001 p. 29)[3]

The National Agency for Education has recently also stressed the importance of working actively with democratic values, using the concept of "deliberative democracy" connected to thoughts of John Dewey and Jurgen Haberma's theory about communicative rationality (Englund 2000 p. 5)[4]. The agency has also initiated a lot of local school projects devoted to the subject.

It is of course difficult to estimate how many of those prescribed central statements are internalized in local every-day school work and by the pupils, and what impact the compulsory school has had on today's peacekeeper, but since the heritage of social movements and the emergence of popular adult education are some of the most important socialization agents in Swedish society, it is reasonable to assume that some of the values and principles communicated through education contributed to the formation of a common value-system.

"Cultural dimensions" in some countries according to Hofstede

Geert Hofstede's taxonomy of cultural dimensions[5] has been criticized by social anthropologists, but has also been used most often in cross-cultural research on leadership.

Examples of cultural Dimensions[6]

Power Distance The extent to which people accept differences in power and status among themselves. In a high power distance culture, leaders have more authority, they are entitled to special rights and privileges, they are less accessible, and they are not expected to share power with subordinates.

High: Russia, China, Philippines, Mexico, Venezuela, India. *Medium:* Netherlands, Italy, Pakistan, Japan, Spain, Greece. *Low:* Israel, Austria, Denmark, England, New Zealand, United States – also rest of the Nordic countries, I believe.

---

3 Johansson, Eva *The UNknown soldier – A portrait of the Swedish peacekeeper at the threshold of the 21st century.* Doctoral thesis, Karlstad University Studies 2001:21

4 Englund, Tomas:*Deliberativa samtal som värdegrund – historiska perspektiv och aktuella förutsättningar* Skolverket 2000

5 See e.g. Hofstede, Geert & Gert: *Cultures and Organizations – Software of the Mind.* Mc Graw-Hill 2005

6 Yukl, Gary: *Leadership in Organizations* Fifth edition Prentice Hall 2002

Individualism The extent to which the needs and autonomy of individuals are more important than the collective needs of the work unit or society. In individualistic cultures, people are identified more by their own achievement than by their group membership or contributions to collective success, and individual rights are more important than social responsibilities.

High: United States, Netherlands, England, Australia, Canada, Belgium – Fairly high also in Sweden. *Medium:* Russia, Japan, Austria, Israel, Spain, India. *Low:* China, Indonesia, Thailand, Pakistan, Hong Kong, Venezuela.

Uncertainty Avoidance The extent to which people feel comfortable with ambiguous situations and inability to predict future events. In cultures with high avoidance of uncertainty there is more fear of the unknown, security and stability are more important, conflict is avoided, plans and forecasts are more valued, and there is more emphasis on formal rules and regulations.

High: Japan, France, Russia, Argentina, Spain, Belgium. *Medium:* China, Netherlands, Switzerland, Pakistan, Taiwan, Finland. *Low:* Singapore, Hong Kong, Denmark, England, Sweden, United States.

Gender Egalitarianism The extent to which men and women receive equal treatment, and both masculine and feminine attributes are considered important and desirable. In cultures with high gender egalitarianism, sex roles are not clearly differentiated, jobs are not segregated by gender, and attributes such as compassion, empathy, and intuition are as important as assertiveness, competitiveness and objective rationality.

High: Denmark, Norway, Finland, Sweden, Netherlands, Chile. *Medium:* Canada, Indonesia, Israel, France, India, China. *Low:* Japan, Austria, Italy, Mexico, Venezuela, Switzerland.

Long-term orientation Perseverance, sustained efforts toward slow results, thrift being sparing with resources, respect for circumstances, concern with personal adaptiveness, willingness to subordinate oneself for a purpose: China, Taiwan, Japan, Vietnam, Korea.

Short-term orientation Efforts should produce quick results, social pressure toward spending, concern with social status and obligations: Nigeria, Spain, Philippines.

According to Hofstede's taxonomy of cultural dimensions, Swedes, together with other Nordic people, are more "gender egalitarian" which could be ethically important in peacekeeping operations, taking into account that women and children are the most vulnerable in insecure societies. Fairly low power distance, to some degree, due to the heritage of popular movement.

Some problematic ethical aspects of young Swedes of today and tomorrow

So far so good, but there are also some more problematic aspects. Sweden has – up till now – an army based upon conscription, but as in most European countries, the percentage of those in an age cohort who actually do their military service has decreased down to 30 % of the male population to be drafted.

"According to a government-commissioned report (SOU 2000, p 21)[7] *a picture emerges that points to significant differences in attitudes among young people who are interested into carrying out their compulsory military service compared to those who are not. Compared to the average group of young people in Sweden, the group of 'those willing to perform military service' held, to a greater extent, the attitude that Sweden should accept fewer refugees and believed that questions concerning law and order were important. Further, those willing to be conscripted were less likely to support the idea that democracy is the best form of government in all situations, attitudes not always in harmony with the attitudes and values generally associated with peace support operations."* (Johansson 2001 p. 24)[8]

According to Hofstede's "cultural dimensions", most western countries are fairly highly rated at the "Individualism" dimension. Surveys of younger Swedes, made by "Kairos Future"[9], shows an increasing individualism from youngsters born after 1980. This "me-we" generation often regards its own interests in the first place and secondly the interests of its own group. They are to a great extent occupied with identity construction and marketing their own "trade mark", and when things are not interesting or fun any longer – they quit.

Swedish politicians have high ambitions regarding Sweden's participation in international missions. But are we also going to get the best of an age group to sign on for military service and international missions in the future?

"With a system that is dependent on individuals' willingness to participate in peace support operations there will always be variations in the number of applicants over time. The system becomes vulnerable to different influences, including for instance, changes on the labour market. Other circumstances that affect the number of applicants are the character of the operation and how media describe it." (Johansson 2001 p. 26-27)

What happens if the intention of the educational system to foster democratic and moral citizens loosens its grip on the coming generation, in favour of a "me-we" individualistic approach to life? If a majority of those, willing to conscript for military service and international missions are not fully devoted to democratic

7 Total Defence service, Official Reports of the Swedish Government, Stockholm 2000
8 Johansson, Eva *ibid.*
9 *http://www.kairosfuture.com/en/international/projects/globalyouth*

and humanistic values? If they are influenced by the "macho ethics" which belong more to situations where you should beat the enemy trying to invade your country? And at the same time, "we are in need of military ethics different from those of the ages of World Wars and the Cold War" (Toiskallio 2007 p. 16)[10]. To meet those possible situations I think we indeed are in need of further developing the ethical education in the military force.

Ethical education in the Swedish Armed Forces

In the pamphlet **"*Here and Now*"**, the Swedish Supreme Commander has formulated the fundamental values of the organization in three "keywords": *Openness* (Cooperation, honesty, showing confidence), *Results* (creative, deliver, be clear) and *Responsibility* (give, take, demand). "Our point of departure is self-evident: All human beings are equal in dignity and rights, respect for every individual, good leadership" (my translation) – A somewhat feeble impression, in my opinion.

As I see it, the ethos and ethics of the Swedish Armed Forces could develop into more structured, coherent and clearer prescriptions as a more elaborated ground for ethical development. – Still, according to my interviews with Swedish officers in Kosovo, those fundamental values formulated by the Swedish Supreme Commander, have had some substantial impact on the ethical awareness among Swedish soldiers.

A comparison between Sweden and the US shows big differences between the Swedish *Openness, Results* and *Responsibility* and the US "Warrior ethos" (always place the mission first, never accept defeat, never quit and never leave a fallen comrade), they have one thing in common though: It is all intrinsically about the own force.

During the military training and education, there are only very limited discussions and training regarding those ethical/cultural aspects of international missions. The articulation and discussion of specific *military* ethos and ethics suited for international missions only occurs to a very minor degree. For example, are 'humanism' and 'democracy' supposed to be central components in the fundamental values for the military force? But the deeper meaning of those complex concepts are seldom reflected upon and made articulate for open discussions.

10 Toiskallio, Jarmo: Edifying Military Ethics. pp 11-27 in Toiskallio (Ed): *Ethical Education in the Military – What, How and Why in the 21st Century.* National Defence University Finland 2007

In a previous article[11] I cited Dr. and Lieutenant Colonel John Mark Mattox who claimed that central to the traditional US Army ethic is the idea of army officers as the defenders of the defenceless. The meaning of this becomes very different whether we mean the defenceless in my own country – or in "the Global Village". That goes for example for Same villages in northern Scandinavia, those few Serbs still left in Kosovo, civilians in Lebanon, Palestine, Iraq etc. Defence of the defenceless in that global meaning comes close to focusing on 'human security' instead of 'state security'.

"'Freedom from fear' and 'freedom from want' have become the catch phrases of an approach to security called human security. Often referred to as 'people-centred security' or 'security with a human face', human security places human beings – rather than states – at the focal point of security considerations. Human security emphasizes the complex relationships and often-ignored linkages between disarmament, human rights and development. Today all security discussions demand incorporation of the human dimension."[12] (Talerud, 2007).

I think, however, that this must be articulated, discussed and implemented as an ongoing conscious process to a much higher degree than in the present situation, and especially according to the very difficult process of peace building and "war amongst the people". This cannot be forced upon the single officer or soldier by codes or decree from above. It must develop from the inside of every soldier.

"A recurrent theme of personal military narratives deals with showing the right 'attitude' or, more specifically, a 'professional attitude' expressed not only through professional expertise and basic soldiering skills but also through the will to improve oneself continually. 'Professionalism' thus embraces the full breath of an individual's character and moral outlook – i.e. the values he represents and exemplifies in the way he presents himself, in this context, 'professionalism' is a military formative ideal,"(Nørgaard & Holsting, 2006)[13].

This "formative ideal", including as I see it, the concept of "Human security", should be integrated in every soldier and officer through a process of lifelong self-education or "self-bildung". At the 8th International Conference on Military Pedagogy in June 2007, Debora Sandler emphasised the importance of the ongoing process of ethical matters; I agree and think we have to learn from other forces when developing our ethical military education. Maybe we also have something to

---

11 Talerud, Bo: Ethos in and Ethics for today's armed forces in the western societies. pp 63-84 in Toiskallio (Ed) *Ethical Education in the Military – What, How and Why in the 21st Century* National Defence University, Finland 2007

12 United Nations Institute for Disarmament Research. *http://www.unidir.org/html/en/human_security.html*

13 Nørgaard, Katerine & Holsting, Vilhelm S: *International Operations in Focus* p. 65 Försvarsakademiet (Denmark) 2006

learn from the Nordic movement of popular education during the 20th century. We do have some interesting educational challenges in this ethical development – at least according to the Swedish Armed Forces, most likely in other forces too.

References

Bron, A. & Talerud, B. (2004). Co-existence or Convergence of General and Vocational Education in Sweden. In E. Lechner & F. Pöggeler (Eds): *Allgemeinbildung und Berufsbildung – Konkurrenz und Kongruenz der Konzepte im Europa des 20. Jahrhunderts*. Frankfurt am Main: Peter Lang.

Englund, T. (2000) *Deliberativa samtal som värdegrund – historiska perspektiv och aktuella förutsättningar.* Stockholm: Skolverket.

Hofstede, G. & Hofstede, G. J. (2005). *Cultures and Organizations – Software of the Mind.* New York: Mc Graw-Hill.

Johansson, E. (2001). *The UNknown soldier – A portrait of the Swedish peacekeeper at the threshold of the 21st century.* Doctoral thesis, Karlstad University Studies.

Kairos Future. [On-line]. Available: *http://www.kairosfuture.com/en/internationalprojects/globalyyouth.*

Nørgaard, K. & Holsting, Vilhelm S (2006). *International Operations in Focus* Försvarsakademiet (Denmark).

Talerud, B. (1985). *Cultural Pedagogy in a Technological Society – On the possibilities of cultural development through pedagogical activities*. Doctoral thesis Department of Education, University of Stockholm.

Talerud, B. (2007). Ethos in and Ethics for today's armed forces in the western societies. pp 63-84 in Toiskallio (Ed) *Ethical Education in the Military – What, How and Why in the 21st Century.* National Defence University, Finland.

Toiskallio, J. (2007). Edifying Military Ethics. In J. Toiskallio (Ed): *Ethical Education in the Military – What, How and Why in the 21st Century.* National Defence University Finland.

United Nations Institute for Disarmament Research. [On-line] Available: *http://www.unidir.org/html/en/human_security.html.*

Yukl, G. (2002). *Leadership in Organizations.* Fifth edition. New Jersey: Prentice Hall.

Harold Simpson

Designing a Defence Studies Curriculum

A soldier without political education is no more than a potential criminal[1]
(Captain Thomas Sankara, former President of Burkina Faso, assassinated during a military coup d'etat, 1987)

Introduction

The International Military Advisory and Training Team (IMATT) has been in Sierra Leone since the end of the civil war in that country since 2001. Despite the "International" in its title 90% of the IMATT membership is drawn from the UK Armed Forces and it is under British political and military leadership. Since its inception the IMATT is acknowledged to have made steady progress in retraining the Republic of Sierra Leone Armed Forces (RSLAF) although the RSLAF continues to experience serious logistical, resource and financial problems.

Understandably, the initial main effort of the IMATT was to concentrate on the individual military training of soldiers and officers. It was very much a "back to basics" approach to put right the fundamental building blocks of the RSLAF. With the heavy emphasis on initial military skills training there was no immediately urgent imperative to introduce a through career long-term developmental educational curriculum. Indeed at the time the civil war ended it would have been premature to do so because the RSLAF was in considerable disarray and would not have been able to identify a coherent educational requirement to be met for either officers or soldiers. Demobilisation and disarmament of those leaving and other fundamental organisational problems needed fixing first. But by 2005, after 4 years of steady progress, the more stable condition of the RSLAF made it timely to design and introduce an overarching scheme of progressive mandatory education for all soldiers and officers of the RSLAF. There were strong political imperatives as well, which will be explained.

1 Sankara, T. (1983). *Discours d'orientation politique,* quoted by Jaffre in Le Monde Diplomatique, Paris, November 2007.

Aim

The aim of this article is to explain the principles of the RSLAF educational scheme, introduced in 2005 that has become known as Peace Education and Training (PET), the rationale for introducing such a scheme and the pedagogical methodology employed to design the curricula. It will also enunciate some "lessons identified" that may be applicable to advisory and training missions in other theatres where the British Armed Forces are engaged in post conflict stabilisation and nation building.

Historical background

Since independence from the UK in 1961 the Sierra Leone Armed Forces (known variously as the Royal Sierra Leone Military Force and Sierra Leone Army until being renamed RSLAF in 2002) had intervened regularly in politics staging periodic coups d'etat although this was a common pattern in post independence Africa. It was the 10-year civil war from 1991-2001 that reduced civil-military relations in Sierra Leone to an all time low with both civil society as a whole and the new democratic government elected in 2002 viewing the Armed Forces with profound dislike and distrust. Government and civilians alike had abundant justification for loathing the Armed Forces; during the civil war the Armed Forces had committed numerous human rights violations against the civilian population and in this respect were considered to be only marginally less culpable than the infamous rebel group, the Revolutionary United Front (RUF). Indeed for much of the civil war the Armed Forces and the RUF colluded in robbing and oppressing the civilian population, coining the word "sobel", an individual who was a soldier by day and a rebel by night. Looting, raping, killing and tens of thousands of limb amputations were the Standard Operating Procedures of the RUF and the sobels. The nadir of the Armed Forces' reputation came in 1997 when a group of middle ranking officers led by Major Johnny Paul Koroma (a Sandhurst graduate incidentally) overthrew both their own High Command and the newly elected civilian democratic government of President Kabbah. Kabbah fled for his life to Guinea and was eventually restored to elected power in 2002 following a series of foreign interventions by Nigeria, the UK and the UN. Having seized power in May 1997 Koroma's junta renamed itself the Armed Forces Revolutionary Council (AFRC) and invited the RUF to join them in sharing power. For 7 terrible months a reign of terror was inflicted on Freetown and much of the countryside by the AFRC and RUF. The reputation and credibility of the Armed Forces was never to recover from their collaboration with the rebels in the late 1990s, especially when they jointly resisted Nigerian forces trying to restore President Kabbah to power. When the final ceasefire eventually came into force in 2001 the reputation of the Armed Forces could hardly have been lower. Both the newly restored government of President Kabbah and the population of Sierra Leone regarded the Armed Forces with distrust and suspicion.

The initial requirement

Once the decision had been made in principle by the newly restored Kabbah government not to disband the Armed Forces altogether the immediately convenient option was to rely on external agencies such as the United Nations Mission in Sierra Leone (UNAMSIL) and IMATT to control the RSLAF and prevent future coups d'etat. But with the phasing out of UNAMSIL forces by the end of 2005 and the planned downsizing of IMATT in the run-up to 2010 reliance on foreign agencies could never provide a permanent solution. The government and its foreign allies realised that a fundamental re-education in democratic values of the RSLAF was the only viable long-term option. If the RSLAF at all rank levels could be persuaded to respect the Rule of Law, the democratic government and the human rights of the people then the likelihood of future coups d'etat and civil wars could be reduced if not eliminated. But this would require all members of the organisation to undergo fundamental attitudinal and cultural changes, arguably a more complex task than conventional military skills training. The strategic political intent having been articulated by the government and Deputy Defence Minister the challenge was taken up in earnest by the RSLAF Directorate of Defence Training Education and Recruiting (DDTER) in early 2005. Despite the grand sounding title the DDTER consisted of only 5 RSLAF officers and 2 IMATT officers who were a British Army Education Officer of lieutenant colonel rank and a Canadian Forces Training Development Officer of major or equivalent rank. Initial scoping of the Sierra Leone government's requirement revealed some very fundamental questions to be addressed, principally…

What content should be put into any new educational curriculum?

Initial analysis rapidly showed that copying and/or adapting the British Army's professional development courses or indeed any other foreign curricula would produce a product of little or no relevance to the local Sierra Leone context. In order to ensure that the local requirement would be met with a relevant solution the DDTER team made an early assumption that the content of any newly developed educational schemes for officers and soldiers of the RSLAF should be identified and developed locally.

Who would identify the curriculum content and carry out the course design work?

In addition to the 2 IMATT officers the DDTER possessed only one RSLAF officer with an educational background capable of carrying out course design work in an academic field such as Defence Studies and developmental education. The other DDTER officers were all generalists comfortable with mainstream training but rather less at home with the concepts of Bloom's Taxonomy of Educational Objectives (more on this later!). Throughout the remainder of the RSLAF only one other suitable officer could be located with sufficient educational expertise to carry out the task of course design in the higher levels of the Cognitive Domain of Bloom's Taxonomy. The harsh reality to be faced was that historically the country's Armed Forces never attracted many well-educated Sierra Leoneans; political and tribal allegiances had always been more important in the recruitment and promotion criteria than intellectual ability. To make matters worse the better educated had largely left the country to work abroad during the civil war and were reluctant to return to an uncertain future in Sierra Leone. In summary, the DDTER team's estimate of troops to task identified the need for more troops early on, specifically experts in curriculum design as well as Subject Matter Experts in Defence Studies and related academic fields. Without a team packing the required intellectual horsepower the danger would be that the whole project would take years to complete and would lose momentum and credibility in the process.

Overseas outsourcing?

The DDTER team did briefly consider possible overseas outsourcing options such as transferring the work back to the UK or Canada using the expertise and resources of existing Training Development organisations. However, even if these organisations had the spare capacity (which was doubtful) to carry out a substantial curriculum design task the problems of being divorced from the local Sierra Leone context persisted. The danger would remain that a series of courses designed outside Sierra Leone by British or Canadian Training Development Officers would prove irrelevant and non transferable once given back to the RSLAF to deliver within Sierra Leone. The same problems would result if an overseas consultant or private military contractor were to be engaged to carry out the work.

This particular solution was in any case almost certainly a non-starter because the RSLAF possessed no funds of its own to engage foreign consultants or contractors and in early 2005 IMATT itself could offer only limited funding (to the tune of several hundred pounds) because of being heavily committed to existing training and infrastructural projects.

Back to a local solution

Once overseas outsourcing had been discounted as an option the DDTER team refocused its attention back to Sierra Leone and examined the possibility of local outsourcing. Scrutiny of local institutions quickly identified the University of Sierra Leone as having the greatest repository of the type of expertise likely to be needed. Within that university a new department soon caught the attention of the DDTER team, namely the Centre for Peace Education and Conflict Resolution Studies. This new faculty had been established with UK government (through the British Council and Department for International Development – DfID) assistance in 2002 and by 2005 had started offering a 3 year bachelor's degree course in Peace and Conflict Studies[2] in addition to numerous local outreach programmes to government departments such as the Ministry of Education as well as local community organisations and civic action groups throughout Sierra Leone. The DDTER team approached the head of the Centre for Peace Studies and Conflict Resolution with an invitation to become a partner with the RSLAF in designing and delivering the new educational curriculum, which as the Sierra Leone government had stated, was a national priority. With the offer of a modest amount of IMATT funding as well as the opportunity to raise its profile in national terms the majority of the staff of the Centre for Peace Studies and Conflict Resolution agreed to participate in both the curriculum design and delivery of the first tranche of pilot courses. This provided sufficient intellectual resources both in terms of subject matter expertise and curriculum design capability. Interestingly the university staff was entirely comfortable with the notion of doing business with the British and Canadian IMATT officers but were wary of and unconvinced by the apparent good intentions voiced by the Sierra Leonean officers they encountered. Throughout the early stages of the partnership the IMATT officers had to act as the essential catalyst, bridging the gulf of mutual suspicion and enmity that existed between the academic community and the RSLAF. Some of the academic staff even

2 Fourah Bay College Handbook (2005), *Bachelor of Social Science with Honours Degree Course in Peace and Conflict Studies,* Freetown: University of Sierra Leone Publications.

feared for their physical safety when entering RSLAF barracks and in the early stages had to be escorted by British and Canadian IMATT personnel.

When questioned about this the academic staff explained, quite reasonably, that they and their families had suffered personally at the hands of the Armed Forces during the civil war. In the same breath, however, they stated that they wanted to play their part in reforming and re-educating the Armed Forces so that the tragedy and brutality of the civil war could never happen again. This idealism and personal motivation was a consistent theme amongst the academic staff who agreed to co-operate with IMATT in the curriculum development project. It is doubtful that a foreign consultant or private military contractor would ever be inspired by such idealistic motives.

Significantly, too, the financial cost of employing 6 university staff for the initial curriculum development and delivering pilots of all the courses over a period of some 6 months would have bought probably at most 1 day of a foreign consultant's or private military contractor's time. By using a local institution as a partner local salary rates could be paid, which in simple cash terms were approximately one fiftieth of the international "going rate" charged by western expatriate consultants and companies. By "going local" far better value for money was achieved right at the outset as well as the likelihood of a far more contextualised and relevant product being developed.

More challenges

Once the local partner had been identified and a basis for co-operation agreed the fundamental problems of identifying content and designing new curricula persisted. The Centre for Peace Studies and Conflict Resolution offered a 3 year degree course covering all aspects of citizenship, human rights, Law of Armed Conflict, International Humanitarian Law, democratic governance, national and international institutions for resolving conflict as well as civil-military relations, social justice, women's rights, children's rights, tribal rights, the environment and a plethora of other topics most of which would be of limited relevance to the RSLAF. Altogether some 500 topics (each roughly equating to an Educational Objective (EO) or Training Objective (TO) in British Army parlance) comprised the degree course that was designed to take 3 years. Apart from being too long and too complex for the needs of the RSLAF, few officers and even fewer soldiers could ever be released from their normal duties for the length of time it would take to complete the degree course. And yet amongst the vast array of subject matter in the degree course it was tantalisingly apparent that somewhere in amongst the

500 or so syllabus objectives could be found the basis of the curriculum so badly needed by the RSLAF.

In order to help identify the most relevant subject matter from the degree course the DDTER team conducted a series of focus groups where a selection of RSLAF officers were asked to peruse all the objectives of the degree course and allocate priorities to each topic in terms of their perception of their military role within the new democratic Sierra Leone.

This process was supervised and quality controlled by the IMATT DDTER officers and the academic staff from the University of Sierra Leone. At the end of the focus groups just over 100 objectives out of the original 500 were identified as being overtly relevant to the military context. Some progress had therefore been achieved but it reality the DDTER team felt it was still a long way from its stated goal of a comprehensive overarching curriculum covering all rank levels for both soldiers and officers. In outline the problems persisting at this stage were as follows:

The focus groups had involved officers of mainly field rank and above. Therefore the objectives identified in the focus groups might be deemed as a suitable basis for proceeding with design of a curriculum for officers of major and above but probably less applicable to the needs of the junior officers and the non-commissioned ranks.

The most thorny problem was the realisation that the focus groups process had really just produced a slimmed down university degree. Although the volume of subject matter had been slimmed down to approximately 20% of the original bachelors degree (100 topics objectives from the total of 500) the intellectual level at which the subject matter was pitched remained that of a university degree course. Intellectually, it would probably be achievable by the most academically gifted element of the officer corps but not by the remaining 95% of the RSLAF.

In summary, the DDTER team felt that a broadly relevant curriculum pitched at the appropriate intellectual level had been identified for the upper echelons of the officer corps but not for the more junior officers or any of the soldiers. The DDTER team made the assumption that ultimately the overall subject matter required for junior officers and soldiers should be similar to that taught to their commanders and other senior officers but would need to be simplified and adjusted according to 2 main criteria.

The first criterion is represented by the intellectual/educational capacity of the differing rank populations. In other words, what intellectual level could a private soldier, Junior NCO, Senior NCO, Warrant Officer or junior officer be expected to cope with? There would be little point pitching the new curriculum at too high a level and at the same time too simplistic a curriculum would represent a wasted opportunity and a poor learning experience for those on the receiving end.

Secondly, what did the job roles of the differing rank populations actually require in terms of training and education to perform their duties and to what intellectual level? Obviously senior officers would be required to analyse and conceptualise at the highest intellectual level whilst junior officers, Warrant Officers and Senior NCOs would be more focused on co-ordination and implementation and private soldiers and Junior NCOs on interpreting instructions correctly and obeying orders effectively.

Job Role Analysis

Once the DDTER team realised that a curriculum of differing intellectual levels tailored to the different rank levels was unavoidable it was decided to carry out a rank related job role analysis. Through such job analysis the generic intellectual capabilities required by each rank group could be identified and a tiered curriculum designed for matching at the appropriate academic level to each rank group. This process would have been substantial to undertake for a large and complex organisation such as the British Army but was relatively straightforward for a small, (roughly 10,000 total), low technology, and infantry based force such as the RSLAF. The chosen tool for the job role analysis was Bloom's Taxonomy of educational objectives[3], which will be explained in the ensuing paragraphs. This mechanism had a number of key advantages; it has been tried and tested over decades throughout the educational world and, more specifically, it was readily understood and accepted by the British and Canadian IMATT officers, the RSLAF and the University of Sierra Leone academics.

Bloom's taxonomy of educational objectives

Between 1949 and 1953 some 40 top educationalists met at a series of conferences in the USA, under the chairmanship of Benjamin Bloom of the University of Chicago. Their aim was to produce an all-encompassing, universal framework for the design and assessment of learning. The end result was a 3 way splitting of learning into intellectual knowledge, physical skills and attitudes (the so called Cognitive, Psychomotor and Affective Domains).

Each of these 3 areas of learning was further sub-divided into a series of hierarchical objectives with a comprehensive overarching framework demonstrating

3 Krathwohl, D. and Anderson, L. (Eds), (2001), *A Taxonomy for Learning, Teaching and Assessing,* New York, Longman.

the inter relationships between the objectives and, in particular, how the mastery of the higher order objectives would always be dependant on the achievement of the lower level objectives in the correct sequence. By way of explanation a summary of the key stages of the Cognitive Domain is shown in the diagram below, arranged in the correct pedagogical sequence. The key principle is that of sequencing and dependency – the learner must start at box 1 and work upwards in strict numerical order to achieve anywhere up to box 6 and its subsets which represent the highest level of intellectual activity. The leaner may of course not progress all the way to box 6 but that is not the fundamental point. The fundamental point is that box 1 represents the simplest level of intellectual learning and box 6 the highest and it is impossible to move upwards from box 1 without progressing through boxes 2,3,4,5 in the correct numerical sequence.

Put simply it is not possible to analyse (box 4) anything unless knowledge (box 1), understanding (box 2), and application (box 3) have been successfully mastered first. By differentiating the intellectual levels of the Cognitive Domain into distinct "layers" Bloom and his colleagues had created a generic educational pedagogy that can be applied to almost any learning situation and can be relied upon to bring an instant intellectual rigour and structure to the curriculum that is to be developed.

Interestingly the Psychomotor (Physical Skills) and Affective (Attitudinal) Domains are also divided up into their own respective numerically sequenced hierarchical taxonomies of objectives. It is possible to use the Psychomotor Domain to assist in designing learning strategies for the acquisition of physical skills in the same way that the Cognitive Domain facilitates the acquisition of intellectual skills. It is also possible, although controversial, to employ the Affective Domain to assess and develop attitudes. Regrettably it is not the aim of this article to explore the Psychomotor and Affective Domains but it is worth mentioning that they exist and may have useful applications in a variety of military and civilian contexts.

Applying the Taxonomy to Job Role Analysis.

Using another series of focus groups the DDTER team matched the intellectual levels of Bloom's Taxonomy by asking the different groupings what they did in their daily jobs in terms of boxes 1 to 6 of the Cognitive Domain. Senior officers in the MOD replied that much of their work revolved around evaluating situations, analysing problems and creating original solutions – levels 4, 5 and 6 of the Cognitive Domain. Junior officers, Warrant Officers and Senior Non Commissioned Officers in the Field Army were found to carry out job roles revolving around executing and implementing fairly standardised procedures in given situations (level 3 of the Cognitive Domain), with only a limited requirement to analyse (level 4) and no evaluation or creating of original solutions (levels 5 and 6).

The consensus was that senior officers would carry out these functions on behalf of everyone else and cascade the results down to their subordinates. Sierra Leone is, after all, a far more traditional and hierarchical society than the UK with more deference to authority and a marked unwillingness to encourage too much initiative by those lower down in the Chain of Command. The private soldiers and Junior Non Commissioned Officers for their parts needed primarily to understand and interpret written and oral instructions correctly (levels 1 and 2 of the Cognitive Domain) so that when directed by Senior Non Commissioned Officers, Warrant Officers and officers they could carry out their orders efficiently and reliably. Even the Junior Non Commissioned Officers had very little genuine responsibility for implementing procedures or carrying out the leadership function on their own because the expectation was that their superiors would give them the necessary instructions. Junior Non Commissioned Officers in the RSLAF are essentially low-level supervisors who are given limited responsibility to assist their Senior Non Commissioned Officers, Warrant Officers and officers in carrying-out group tasks. They are not expected to display anywhere near the same degree of leadership that the British Army expects of its Junior Non Commissioned Officers.

Results in the Job Role Analysis measured against the cognitive domain of bloom's taxonomy

In outline the results of the Job Role Analysis looked like this:

Senior Officers. Typically operating in their jobs mainly at Levels 4 and 5 – Analysis and Evaluation. On occasions at Level 6 – creating and synthesising wholly new products, such as designing a new doctrine for the RSLAF.

Junior Officers/Warrant Officers/Senior Non Commissioned Officers. Typically operating in their jobs mainly at Level 3, applying, executing and implementing the right procedure in the right given situation. Officers also reported a significant analytical element in their work, equating to Level 4 of the Cognitive Domain.

Private Soldiers and Junior Non Commissioned Officers. Typically operating in their jobs at Levels 1 and 2, rarely any higher.

Recruits in training. No job roles as such to identify but the likely intellectual need was identified as a simple knowledge requirement (Level 1 only) that would be reinforced on arrival in unit with a package to bring the individuals up to Level 2.

Curriculum design for the pilot Peace Education and Training (PET) courses

Having identified the appropriate levels of the Cognitive Domain for each rank grouping the next step was to "map" them against the 100 or so objectives from the bachelor's degree in Peace Education and Conflict Resolution that had been initially identified by the early focus groups as being relevant to the RSLAF. Most of the original degree course was pitched mainly at Levels 4,5 and 6, which were the same levels that had been identified for the senior officers during the Job Role Analysis. Therefore the senior officers ended up with cherry picked elements of a slimmed down degree course. The junior officers, Warrant Officers and Senior Non Commissioned Officers had their levels of learning the 100 or so objectives reduced from the predominantly 4, 5 and 6 of the original degree course to Levels 1,2 and 3 with an element of Level 4 (Analysis) for officers. This again precisely matched the findings of the Job Role Analysis. Finally the private soldiers and Junior Non Commissioned Officers had their levels of learning the 100 or so objectives reduced to Level 1 and Level 2 in line with the findings of the Job Role Analysis.

By reducing the intellectual levels from the original Level 6 of the bachelors degree course the amount of time taken to teach each objective was much reduced as well as ensuring that the new curriculum would be suitable for and achievable by the target populations represented by each rank grouping. For example the Law of Armed Conflict can be taught to Level 1 by rote as a simple exercise in memorising the key facts in less than 1 hour. If simple memory recall (Level 1) is the objective rather than understanding (Level 2) then the objective can be achieved very quickly. If Level 2 (understanding the principles in addition to memorising key facts) becomes the objective, then the lesson will take longer. The British Army's now obsolete Law of Armed Conflict Individual Training Objective (ITD) was a good example of a learning event that was pitched at Level 2 in that the learners had to demonstrate that they had memorised the key facts and understood the underlying principles but were not required to apply what had been learned in any type of scenario, simulated or otherwise. If the ITD had included a requirement to go to Level 3 (implementing and executing) by testing students in simulated scenarios to determine whether they could apply the correct procedures successfully then the ITD would have increased in length from roughly one hour to probably several hours or even a whole day. A daylong seminar of some sort would be the likely outcome. Finally, if learning the Laws of Armed Conflict were to be designed to Levels 4, 5 or 6 (analysis, evaluation and creating new theories about the legality of Armed Conflict) it would take days or weeks to complete. Indeed a module of a law degree would effectively be the outcome if the learning was developed all the way up to Level 6. Thus by altering the Taxonomy Level applied to each learning objective the complexity and time taken could be drastically reduced or adjusted. To take the Law of Armed Conflict as an example, learning pitched at Level 1 might take 40 minutes, Level 2 might take 1 and a half hours, Level 3 might take half a day, Levels 4 and 5 several days and Level 6 a week or several weeks.

Having identified the correct levels of Bloom's Taxonomy for each of the 4 rank groupings and by applying these levels to each educational objective it was relatively straightforward for the DDTER and University of Sierra Leone team to create draft Instructional Specifications, the essential basis for running pilot courses. By April 2005 the first of the pilot courses was ready to be trialled.

The launch of the pilot pet courses initial reactions

The programme for the PET pilot courses can be seen below. The courses were taught mainly by University of Sierra Leone but also by RSLAF officers wherever there was a suitably qualified officer available. Of passing interest is the legacy module in the local language (Krio or Creole), which was a transitional concession to the significant minority of semi-literate soldiers in the RSLAF who were recruited during the height of the civil war when the Army was desperate for manpower and no educational entrance criteria were applied. This situation had changed completely by 2004 (with a new set of recruiting criteria requiring 5 West African Examining Board passes at Form 5 level just to join the infantry) but the legacy soldiers needed to receive the same training as the newly joined. Not only did many of the legacy soldiers possess limited literacy but neither were they functional in Standard English. The pragmatic solution was to develop a Krio version of the private soldier/Junior Non Commissioned Officer in unit module with only a minimal written element. As the less educated wartime soldiers are progressively replaced by the new generation of well-educated recruits the Krio legacy module will be phased out. All other PET modules are written and taught in Standard English.

All pilot modules were enthusiastically received both by those being taught and those carrying out the teaching. Inevitably further developmental work will have to continue during 2006 and 2007 on all the PET modules but the pilots carried out in 2005 proved the concept. The intellectual levels were generally correct for the 4 rank groupings and the subject matter endorsed as being not only relevant but also long overdue. An additional benefit was that a rapprochement began to develop between the RSLAF and the academic community after many years of hostility and mistrust. Both sides saw what the other had to offer and, more importantly, how much more could be achieved for Sierra Leone by working on co-operation with each other. Significantly, other West African countries (including at least one Francophone country in the region) showed an interest in the RSLAF PET scheme and expressed a wish to develop something similar. At the time of its launch the RSLAF scheme was the only one of its kind in the West African sub region. For once Sierra Leone, accustomed to being regarded as the most hopeless country in West Africa, had actually led the way in a significant area and produced a novel product that others wished to emulate. This achievement could not fail to have a positive effect on both RSLAF morale and the morale of the academic community.

Conclusion

The RSLAF PET scheme was built on an unconventional approach to curriculum development and course design but nevertheless managed to produce a "fit for purpose" product in a short space of time at minimal cost. More importantly, the PET scheme relies largely on local expertise and capability, which augurs well for its long-term sustainability after IMATT has downsized and departed from Sierra Leone.

All RSLAF personnel from recruit to senior officer now receive a mandatory educational curriculum appropriate to their rank level and commensurate with their intellectual ability. The common themes covered in all the core curricula are conflict prevention and management, security concepts and studies, arms control and proliferation, peace-keeping operations, human rights, international humanitarian law, the law of Armed Conflict, democracy, governance, civilian control of the Armed Forces, citizenship and an introduction to African regional organisations.

The RSLAF and other local Sierra Leonean institutions have successfully met the requirement set by the Government of Sierra Leone, although with significant help from British and Canadian IMATT officers in kick-starting the process. The hope is that having successfully launched PET, the scheme will achieve enough momentum of its own to be self-sustaining.

Lessons identified

The following lessons identified may be relevant for other nation building and post conflict stabilisation missions being undertaken by the British Armed Forces and those of other Western countries. At the time of writing (mid 2006) British and other Western missions in countries such as Iraq, Afghanistan and East Timor spring to mind as well as the smaller Military Advisory and Training Teams operating in many parts of Africa, Asia and the Middle East.

Don't overlook local expertise, although it may not be immediately apparent especially in a society emerging from a recent conflict or still in the throes of a civil conflict. You may have to do some searching in unlikely quarters in order to find the local expertise you need but your persistence may well pay dividends.

Local expertise is almost always cheaper and often more effective than imported expertise brought in by foreign military organisations, aid agencies, contractors, consultants and NGOs. Employing locals will also provide much needed jobs and income and boost the local economy. Employing foreign expatriates of any type does little to help local employment prospects and most of their salaries will in any case go back their country of origin.

Basing a project on local expertise will strengthen its chances of long term sustainability after the foreign intervention force has gone home. A project that is overly reliant on foreign expertise (military or civil) will almost certainly collapse as soon as the foreign presence is withdrawn.

If the intervention is in a post conflict society or a society where conflict is still occurring you may find that local institutions do not wish to co-operate with each other. In particular there may be a schism between the military and civil society. For this reason you may encounter hostility or a reluctance to become involved with the military and other security forces. As the foreign intervention force you have more chance than any other agency of overcoming suspicion and distrust. You have the opportunity to act as the catalyst to get local institutions working with each other instead of against each other. No one else can probably do this.

It will almost always be necessary to kick-start a new project with some initial financial investment (to pay the people whom you have engaged, if nothing else) and other types of resources such as access to transport, stationery, and use of IT and reprographic facilities. This will require an initial outlay of money but almost certainly much less money than it would cost to bring in foreign help. Local expertise may be cheaper but is rarely free!

References

Fourah Bay College Handbook (2005). *Bachelor of Social Science with Honours Degree Course in Peace and Conflict Studie.* Freetown: University of Sierra Leone publications.

Krathwohl, D. and Anderson, L. (Eds.) (2001). *A Taxonomy of Learning, Teaching and Assessing.* New York: Longman.

Sankara, T. (1983), *Discours d'orientation politique,* quoted by Jaffre in Le Monde Diplomatique, Paris, November 2007.

Tamir Libel

Late Bloomers – Professionalising Israel Defence Forces[1]

During the 1990s it was argued that the IDF (Israel Defense Forces) had begun shedding its traditional 'Peoples Army' characteristics in favor of professional ones. Evidence supporting this argument included adoption of a selective draft, privatization, and reduced involvement in nonmilitary issues. The current essay will attempt to further ascertain the validity of this conclusion by examining another aspect of military professionalism, e.g. professional military education. The objective will be pursued through an analysis of the changes in junior Army combat officer education.[2] Since changes of this nature require a lengthy process, the current study will begin in 1973 at which time the IDF underwent dramatic changes, following the 1973 Yom Kipur war ; it will end in 2000 which saw significant reforms in the Officer Candidate School and the beginning of a new chapter in the Israeli-Palestinian conflict. The first part of the essay will discuss the theoretical basis and the dynamics that led to the reforms in the education of junior Army combat officers. The second part will explore these changes and examine their impact on significance and on whether or not the IDF has indeed adopted a professional attitude toward officer training.[3]

What is Military Professionalism?

Professions are generally defined as socio-occupational forms of organization characterized by functions and structures, attitudes, forms of knowledge and unique socio-cultural styles (Siegrist, 2001, 12154). While these characteristics may be of equal importance, Snider and Watkins argue that the uniqueness of a profession lies in the application of abstract expert knowledge to specific problems (Snider and Watkins, 2002). Consequently, this expert knowledge and its acquirement are of the utmost importance. Professionalization, i.e. the adoption

1 This essay is based upon an M.A thesis which was written under the supervision of Prof. Stuart A. Cohen at the Department of Political Studies in Bar- Ilan University, Israel.

2 The term Army is an anachronism for the configuration itself is new to the IDF. For its development in the IDF see: Shalom, 1995.

3 The three services educate combat officers separately; this article discusses only the education in the Army.

of professional characteristics, denotes the processes influencing the social and symbolic construction of profession and status. As abstract models, professions are differentiated by abilities and skill-sets, professional knowledge and the existence of institutions dedicated to imparting this knowledge (Siegrist, 2001, 12154). In affect, these institutions play a crucial role in the process of professionalization (Siegrist, 2001, 12156).

The theoretical basis for the current essay is derived from Samuel Huntington's *The Soldier and State* (Huntington, 1957).[4] Huntington argued that only officers who acquire the knowledge and skills necessary for 'management' rather than 'implementation' of violence should be considered professionals (Harries-Jenkins, 1990; Huntington, 1957).[5] In addition, he suggested that military professionalism is uniquely characterized by expertise, responsibility and esprit de corps.[6] This characterization facilitates an investigation of the role played by military education in the development and preservation of military professionalism, as it is education that affords military practitioners the necessary professional and normative knowledge.[7] The structure of the military education system parallels the levels of war: military academy (Tactical Level), command and staff college (Operational Level) and senior military colleges (Strategic Level) (Sarkessian and Taylor, 1975).[8]

An Identity Crisis: 1973-2000

Several changes occurred in Israeli society and its security environment during the research timeframe, which profoundly affected the characteristics of junior Army combat officer education:

4 Regarding the conceptual framework of military professionalism see: Burk 2002. Burk reflects that ‚military' was only recognized as a profession by social scientists in the 1950s.

5 Increased involvement in non-combative military activity resulted in the broadening the definition military professionals. See: Harries- Jenkins, 1990.

6 Huntington, pp.8-10.

7 This conceptual framework was adapted by the author and will be elaborated upon in his forthcoming PhD dissertation.

8 Senior military colleges are known alternately as War Colleges, Defense Colleges or National Security Colleges.

Strategic Environment

Due to the heavy losses suffered among junior Army officers in the 1973 war and the massive force structure expansion that followed, nearly doubling the number of armored brigades, an immediate "infusion" of junior officers was needed. This was accomplished by both reenlisting retired officers and expanding the officer candidate pool. This last reform dramatically transformed the social composition of the officer corps. By the evening of the First Lebanon war (1982), its last conventional war, the IDF was numerically stronger and better equipped (fielding American weaponry) than ever before. However, despite efforts in the contrary domains of command and control, combined arms operations and air-ground cooperation that had plagued the IDF in 1973 continued to do so nearly a decade later.

Later, junior Army combat officers faced new challenges with the onset of the first Intifada in 1987.[9] Though sub-conventional challenges had long characterized Israeli history, their influence over combat doctrine and force structure had been negligible (Cohen, 1994). It was only through the Intifada and the occupation of the Southern Lebanon's Security Zone that the IDF realized that counterinsurgency operations are unique experiences requiring the development and adoption of specific military expertise (Libel, 2007).

Social Changes

These essentially reflected concurrent changes in Israeli civil-military relations (Peri, 2001; Ben- Eliezzer, 2001). Cohen and Suliman argued that the IDF underwent a process of specialization during the 1990s, abandoning 'People's Army' characteristics in favor of professional ones. The new characteristics included adoption of a 'selective draft' policy and the pooling of resources in order to retain essential manpower (e.g. combat and technologically oriented). The latter included salary incentives and financing for higher education (Cohen and Suliman, 1995). This last provision was made possible through a change in the attitude toward aca-

9 The first Intifada demonstrated Israel's vulnerability to sub-conventional threats; the Iraqi missile attack (1991) demonstrated its vulnerability to NBC threats. The latter effected Israeli Security Doctrine and IDF force structure but its influence over the Army was felt primarily by senior officers and is therefore outside the purview of the current article. See: Cohen, A., Eisenstadt, M. and Bacevish, A. J., 1998.

demic studies, which were no longer considered a reward for excellence but rather a means of enriching professional knowledge and skills.[10]

The Training of Officers, 1973 – 2000

During the research timeframe, junior Army officers were educated through one of the three variations of the Officer Candidate School (Bahad 1). Infantry – a six-month course for infantry, paratrooper and special units; Combat Arms – a three-month course for armor, artillery, combat engineers and anti-aircraft cadets followed by a three-month specialization course; Basic – a three-month course for non-combat cadets, followed by a specialized branch course. The School curriculum had undergone few changes up until the 1990s when it began to raise and expand the levels of knowledge and skills it provided to its cadets.[11] Concomitantly, the school defined four necessary characteristics of command: Professionalism, Control Skills, Values and Leadership (State Comptroller, 1993).[12] An examination of the changes that have occurred in the attitude toward these characteristics will reflect the transformations in officer training.

Professionalism and Control Skills

Formally, admittance to the school had been dependent on passing an entrance examination, a stipulation apparently not always properly enforced. A 1993 study revealed that some cadets had failed their exams in critical subjects (such as navigation) (State Comptroller, 1995). Consequently, advanced courses had to be dropped in favor of basic skills. In 1995 it was determined that parent units would be held responsible for preparing cadets for the entrance exams. An additional impediment to the development of professional knowledge and proficiency was a scarcity of current doctrinal publications. Rising to the challenge, Ground Forces Command published a number of techno-tactical training manuals, which the school was

10 LTC Chen Yitzhaki, Commandant of the Command and Staff College (1993 – 1996), argued in a 2007 interview with the author that the absence of a significant conventional military challenge from 1982 through 2006 forced the IDF to seek substitutes for experience. According to Itzhaki, education was such an alternative.

11 For an historic discussion of the preparation for the course and the training it provided up to the mid 1980's, see: Gal, 1986, 118 – 122.

12 Though an adoption of a military ethical code, especially a formal one, is a prerequisite for the development of military professionalization, it is outside the purview of the current article. Concerning the adoption of formal military ethical code by the IDF see: Eitam. 2003.

required (but failed) to adapt for its teaching purposes (1995). According to the State Comptroller, the length of the officer-education course severely curtailed the school's abilities to offer adequate instruction in regard to both breadth and depth of critical issues (State Comptroller, 1995).

Leadership

Professional authority over the issue of leadership remained undefined until it was delegated to the School of Leadership Development (1994), which was then tasked with developing leadership training in the Officer Candidate School (State Comptroller, 1995). In 1996, the then CO of the school, LTC Itizik Gonen, defined the school's purpose as *mirroring*, or enabling cadets to understand their strengths and weaknesses and offering tools compensating for the latter (After and Halleli, 1996).

Reform efforts were hindered by a failure to define minimal levels of professional and command experience required of team leaders. A radical reform implemented in the school in 1995 replaced the traditional squad and platoon units with 15-man teams. This may have been an attempt to provide cadets with a role model from whom they could learn officer-like behavior. Team leaders were designated 'commander' rather than the traditional 'instructor' in an effort to bridge the conceptual gap between training and command positions. Their responsibilities were broadened with the same objective in mind. However, not all team leaders possessed the necessary levels of professional knowledge and skills (State comptroller, 1995, 905- 906).

Tactical Command College

The establishment of the Tactical Command College in 1999 was preceded by intensive preparation including an exploration of similar institutions abroad and lessons gained from the *Barak* program, which will be discussed shortly.[13] Though the military curriculum was to be taught on an academic level, it was decided that the instructors would all be chosen from among combat unit officers and that the commandant be a Colonel with prospects for senior command. This decision was

13 Interview with LTC (Ret.) Amira Raviv, 2005. Raviv commanded the Military Colleges' Learning and Development Center, and was directly involved in planning for the new college. According to Raviv, the Viennese Maria Theresa academy inspired the new Israeli institution. The assertion concerning the indirect influence of the *Barak* program is being made by the author.

calculated to entice prospective candidates for these positions. These provisions helped thwart opposition to the new school. The program included study towards a BA in military history at the Hebrew University of Jerusalem, intended to expose students to the broader aspects of war waging (Asher, 2003). In addition they enjoyed a multidisciplinary military and security studies program (Kadish, 2004). The students, required to sign on for an additional two years, were experienced lieutenants having already completed two tours and been selected for company command (Kadish, 2004). The military aspect of the curriculum was designed to develop tactical understanding and creativity. Additionally, the school emphasized themes especially relevant to Low Intensity Conflicts such as ethics and the media (Davidi, 2004).

'Barak' Command and Staff Program

The program was born of lessons gained from the First Lebanon War. As a result of the command and control difficulties then encountered it was decided to establish a department within the General Staff's Training Branch dedicated to command and control issues. The lessons learned and the newly formed doctrines were taught in a new officer training program designated '*Barak*' (which translates as "lightening"). Col. Dr. Hanan Shai was charged with establishing the program, assisted by the General Staff Training Branch' LTC Dr. Yosef Hochbaum and Major (Res.) Binyamin Amidror (who was actually a civilian employee of the ministry of defense) and an external advisory team.[14] Speaking of the program Shai remarked: 'the greatest revolution I did in Tzahal… is that command is a profession' (Hornstein, 1994).[15]

Barak was begun as an experiment in 1989. Of the seven or eight teams undergoing the traditional Staff and Command course at the time, two were selected to

14 Shai was involved in the implementation of additional reforms proposed following the 1982 war. Interview with Shai, July 14th, 2005.

15 The team of advisors included Yoav Gelbar and Yehoshaft Harkabi. Col. (Ret.) Gelbar is a leading Israeli military historian. The late Maj. Gen. (Ret.) Harakabi commanded Israeli military intelligence in the 1950s and later became a Security Studies expert in Israel. Gelbar was recruited to the team on the basis of his military rather than academic background. Gelbar, email correspondence with the author, 30.0.07. As head of the General Staff Training Branch, Doron Rubin supported the program. Shai's ideas concerning military training and professionalism were influenced primarily by the Prusso-German experience. Interview with Shai, 2007. The interview was conducted in Hebrew and any errors in translation are the author's.

form the first class of the new program.[16] Table 1 details some of the differences between the traditional course and the new program.

Barak incorporated three military education innovations in the IDF: (1) courses sought to provide students with an academic-level professional military education; (2) it was based on the understanding that imparting this type of knowledge required research oriented instructors; (3) the program provided students with a comprehensive experience cultivating self development including an Officer's Club, attention to proper appearance and bearing, physical fitness, etc. In addition, students were encouraged to continue this process independently; the program imbued students with reading habits through requirements of periodical book reports of military literature.

Barak lasted six years. Though it was never officially disbanded, its staff was gradually pushed (often derisively) aside until its unique character had been neutralized. Under the leadership of a new commandant, LTC Chen Yitzhaki, the school in 1993 began to plan for a new reform. This reorganization was to reflect the challenges graduates would face in the field and therefore introduced various themes hitherto neglected even by *Barak*.

	Traditional Command and Staff Program	**New Program – *Barak***
Military Theory	285 hours (11.4%) emphasis on combat doctrine, staff work and analysis of historic campaigns	512 hours (26/1%), emphasis on combat, organizational, command and staff and control doctrine, analysis of historical campaigns
Exercises (practicing theoretical studies)	630 hours (27.2%), Situation estimates for all possible theaters of operations	280 hours (14.3%), Situation estimates for all possible theaters of operations
Academic studies (such as Political Science)	450 hours (19/4%)	208 hours (10.6%), emphasis on military history and geography, public administration and Zionism
Complementary Military Studies	578 hours (24.9%), emphasis on English language and computer skills, military management, national security, IDF history, physical fitness	539 hours (27.5%), emphasis on English language and computer skills, physical fitness, research and development, technology, IDF history, national security, communications

16 Classes and exercises in the traditional IDF command and staff program were conducted by teams, in a manner similar to the Syndicate system at the historical Staff College of the British Army at Camberley or the American Seminar system at the U.S Army War College.

Independent research, miscellaneous and vacations	391 hours (19.4%)	418 hours (21.3%)
Course duration	2315 hours: 1 week introductory, 3 weeks vacation, 48 weeks study	1957 hours: 2 weeks introduction, 5 weeks vacation, 45 weeks study
Experienced combat officers instructors	Taught military theory and complementary military studies; served as instructors and controllers in war-games and tactical exercises without troops	The position was cancelled
Academic Studies	In one year a student studies 75% of a B.A degree in general history, political science or General Studies[1]	The military program was given an academic accreditation. *Barak* course work was recognized as 50% of the degree requirements.[2]
	Traditional Command and Staff Program	**New Program – *Barak***
Nature of exercises	Themes allotted identical time brackets (be they command and control or logistics) and taught in a technical manner not based on a thorough understanding of doctrine	First semester – instruction of elementary themes. Second semester – case-study analysis and command post exercises. Third semester – division and corps battle management exercises.[3]

Table 1. Differences between the traditional course and the new program

These included preparation for Operational level command, an emphasis on Jointness, management etc (Yitzhaki, 1995).[17]

Was a Professional Model Adopted?

The evidence discussed in this essay points to a duality regarding the professionalization of junior Army combat officer education in the IDF.

On the one hand, the various levels of junior officer education had undergone a process of professionalism: the Officer Candidate School, the establishment of a professional officer education program (*Barak*) and the establishment of the Tactical Command College. It appears that the latter's success was due in part to compromises made in the nature of the education it offered. Its success indicates

17 Though *Barak* did have limited inter-service cooperation, it had attracted much opposition. At that time, the Air Force and Navy possessed parallel command and staff courses comparable to *Barak*. According to Shai these were advanced training courses rather than Command and Staff courses and were ill-equipped to offer students insight into the inter-service aspect of command and control. Interview with Shai, 2007.

that the IDF had internalized the understanding that changes in training reflect not only the professional realm but also on its identity and organizational culture (Sapir, 2006).

On the other hand, the professionalization of the IDF had not kept pace with contemporary reforms being made in Western military establishments; the latter had begun to emphasize combined arms, inter-service and inter-agency cooperation. During the research timeframe the IDF established the Tactical Command College, a military academy offering professional education for tactical level command. However, while similar Western institutions commission officers, the Israeli school was intended to broaden the education of experienced combat officers. From 1989 through 1994 the IDF also maintained the *Barak* program, which provided professional military courses at the Command and Staff level. However, while similar Western institutions offer military education for Operational level commanders, the IDF program only educated officers for higher tactical-level command.[18]

From the research perspective, the current essay has demonstrated that attention must be paid to the role played by military knowledge and professionalism in the transformation the IDF underwent during the past two decades.

References

After, Y. And Halleli, S. (1996). "A Guide to the Perplexed". *Bamachane*, 29, 26- 27. [Hebrew]

Asher, I. (2003). ‹The Art of War and the Theory of Couplehood›, *Ma'ariv* (July 11th, 2003). [Hebrew]

Ben Eliezer, U. (2001). From Military Role- Expansion to Difficulties in Pease Making: The Israel Defence Forces 50 Years On". In: D. Maman, E. Ben- Ari and Z. Rosenhek (2001). (Eds.). *Military, State and Society in Israel*. New Brunswick and London: Transaction Publishers.

Burk, J. (2002). „Expertise, Jurisdiction and legitimacy of the Military Profession". In: L. J. Matthews (Ed.) (2002). *The Future of the Army Profession*. Boston: McGraw Hill: Macgraw- Hill Primis Cutstom Publishing.

Cohen E. A., Eisenstadt, M. and Bacevich, A. J. (1998). *Knives, Tanks and Missiles: Israel's Security Revolution*. Washington, D.C: Washington Institute for Near East Policy.

18 It should be mentioned that the IDF established the Operational Theory Research Institute (OTRI) intended to provide operational level education during the very same year it disbanded *Barak*. However, the Institute's contribution to Operational Level doctrine and officer education remains highly controversial.

Cohen, S. A. (1994). "How Did the Intifada Affect the IDF?". *Conflict Quarterly*, 14/3, 7- 22.

Cohen, S. A. and Suliman, I. (1995). "The Israel Defense Force (IDF): From a 'People's Army' to 'a Professional Force'". *Armed Forces and Society*. 21/2, 237- 254.

Davidi, D. (2004). "The Influence of the Tactical Command College over IDF Officers and Training". *Ma'arachot*. 396, 32- 35. [Hebrew]

Eitam, R. (2003). *The IDF Ethical Code*. MA Thesis, Political Studies Department, Bar Ilan University, Ramat- Gan. [Hebrew]

Gal, R. (1986). *A Portrait of the Israeli Soldier*. New York: Greenwood Press.

Harries-Jenkins, G. (1990). The Concept of Military Professionalism. *Defense Analysis*, 6/2, 117- 130.

Hornstein, G. (1994). A Khaki University. *Bamachane*. 31, 24- 29. [Hebrew]

Kadish, A. (2004). "The Academic Program for IDF Officer- Principles and Objectives". *Ma'arachot*. 396, 30- 31. [Hebrew].

Lev-Tzur, U. (1990). "Operational Exercises and Instructor Experience are the General Tools in Command and Staff College Officer Training". *Chamanit*. 16, 25. [Hebrew].

Libel, T. (2007). *Crossing the Lebanese Swamp: IDF Structural and Doctrinal Adaptations to Counterinsurgency Warfare and its Implications for Al Aqsa Intifada*. Paper presented at the 23rd Annual Conference of the Association for Israeli Studies, Ra'anana, Israel: 11 June 2007.

Peri, Y. (2001). Civil-Military Relations in Israel in Crisis. In: D. Maman, E. Ben-Ari and Z. Rosenhek (2001). (Eds.). *Military, State and Society in Israel*. New Brunswick and London: Transaction Publishers.

Sapir, Reuma. (2006). "Three Prisms on the Construction of Officership in the IDF". *Bein Hazirot*, 5, 35- 43. [Hebrew]

Sarkessian, S. and Taylor, W. J. Jr. (1975). The Case for Civilian Graduate Education for Professional Education. *Armed Forces and Society*, 1/2, 251- 262.

Shalom, D. (1995). *How to Pursue Organizational Innovation? The Establishment of the IDF Ground forces Command*. Tel- Aviv: College of Management. [Hebrew]

Siegrist, H. (2001). *„Professionalization/ Professions in History"*. In: *International Encyclopaedia of the Social and Behavioural Sciences* (Vol. 18, pp.12154- 12165). Oxford: Elsevier.

Snider, D. M. And Watkins, Gayle L. „Introduction". In: L. J. Matthews (Ed.) (2002). *The Future of the Army Profession*. Boston: McGraw Hill: Macgraw-Hill Primis Cutstom Publishing.

State Comptroller (1993). «Officer Cadet Training». *44th Annual Report*. Jerusalem: Governmental Printing House. [Hebrew]

State Comptroller (1995). „Follow Up on Officer Cadet Training. *46th Annual Report*. Jerusalem: Governmental Printing House". [Hebrew]

Yitzhaki, C. (1995). *Changes in the Organization of Military Training and Means of Implementation- A Case Study*, MA thesis, School of Education, Tel Aviv University, Tel- Aviv. [Hebrew]

Juha Mäkinen

The learning and knowledge creating school – Case of the Finnish National Defence College

Abstract

This research focuses on the Finnish National Defence College at the time when the claimed "learning organization-likeness" has been located in the social landscape of the Finnish knowledge society. Having an aim to develop the teaching practices of the Finnish National Defence College, the researcher took a double-hermeneutic stance towards the current teaching practices at the College. In the progressive double-hermeneutic inquiry process five Development Laboratory meetings were arranged at the College in 2004 and 2005, followed by thematic interviews of the current managers of the College. Consequently, prevalent cultural basic assumptions, challenges, problems and key educational principles were identified, offering a fruitful starting point for forthcoming extensive analyses and organizational learning. The results severely challenge the identified "impossibility assumption" of societal guidance by offering some appropriate means for reconsiderations and developmental actions, potentially applicable also in other kinds of institutions and organizations.

A brief introduction to the present educational challenges at the Finnish National Defence College

In the very first Development Laboratory meeting teacher X expressed some of his basic assumptions related to teaching as follows:

Excerpt 1

When focusing on the essence of teaching we have to ask what it really is. Obviously it is *the transmission* of culture from the older generation to the younger generation. From my point of view the question about what will be transmitted and where it will be transmitted from is the most essential question… Consequently, the question of how the teacher arranges his transmission duties in practice is a completely secondary issue... From my point of view these kinds of [developmental] activities are *useless*.

Although the above opinion did not seem to be widely shared by the participating teachers, an important question came to the fore: Could it be possible that the current managers of the Finnish National Defence College (FNDC) and also the teachers of the FNDC share the opinions of the teacher X?

In the thematic interviews the managers of the FNDC understood the concept of learning very differently. Some of them emphasized the fact that the students need some background knowledge and understanding to be able to solve practical problems. It could be said that generally speaking learning was identified to be a combination of learning "old things" and creating something "new". Learning was not considered to be limited to "internalization" or "appropriation" of the given information, but it was also seen to have a functional and practical aspect (e.g. the capability to solve practical problems). So also the managers of the FNDC seemed to be challenging the transmission kind of education, but obviously the teachers of the FNDC are not aware of this fact. Hence the question of systemic guidance of the current teaching practices at the FNDC comes to the fore.

In the case of the FNDC, when starting a developmental process, the question of the tractability of societal systems was actually expressed in the following form:

Excerpt 2

But discussing the change and development of our objectives brings to my mind an idea that there could be such an issue that we cannot guide and affect.

The participating teachers felt that at least they as teachers cannot have an effect on the societal and organizational change processes at the FNDC. Paradoxically, we often tend to neglect the fact that our social and organizational reality consists of individual actors or agents who play a dominant role in the game of life. Giving primacy to the interactionist paradigm (Boudon 1982, Giddens 1979, 1984; cf. e.g. Wenger 1998), it is reasonable to see that without human agency human societies or social systems would plainly not exist. But actors do not create social systems: instead they reproduce or transform them, remaking what is already made in the continuity of praxis (Giddens 1984, 171).

Speaking in a fundamental manner, life is a dialogue among relatively autonomous "embedded *holons*" (Sahtouris 2000, 52; Allee 2003), or agents, actors, human beings, if so preferred, *all of which are critical to the function of the complex social system*.

Actually when asked about the role of the individual teachers in the Bologna process (see Bologna Process), and in the educational development in general, the centrality of the teachers in the process was a shared opinion of the managers. As expected, the teachers were paradoxically unaware of the expectations of the managers and they all seemed to be unaware of the need to reflect on the very essence of social systems.

Our habituated ways to see our current social reality is challenged by figure 1 (Mäkinen 2006).

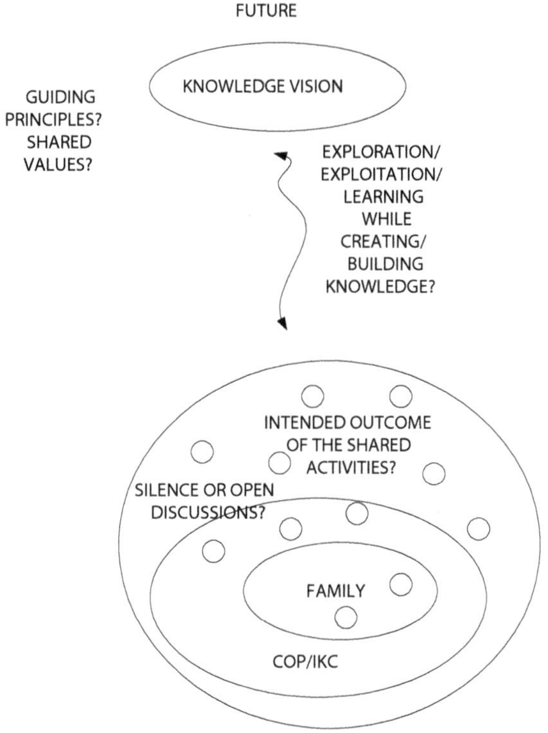

Figure 1. Seeing our social reality from an alternative angle

Figure 1 reminds us of the fact that e.g. in an educational institution, such as the FNDC, relatively many individual teachers (illustrated by the circles in the figure 1) make several influential decisions (e.g. pedagogical; educational) almost on a daily basis. Therefore, it is not appropriate anymore to focus solely on the written curriculum as the one and only object of the ongoing military educational planning processes. Curiously, the layered structure of the curriculum (i.e. written; taught; learned) was already a widely shared interpretation among the managers of the FNDC.

Figure 1 suggests that the attention should be turned towards the shared principles. Here it is assumed and expected that rule-governed interaction among a set of interconnected individuals *can potentially* generate emergent structures (Monge and Contractor 2003, 85, Jackson 2000, 88, Kupers 2001, Anderson 1999, Wheatley 1999), but which ones are the ones currently emphasized and shared at the FNDC?

One main principle often emphasized by the managers was the comparability of the degrees (civilian versus military ones). The principle has been highly valued at the FNDC, aiming to obtain and sustain a university status. The importance of the comparability principle has been increased also through the Bologna process. Despite of this, the degrees of the FNDC have been shifting between comparable and incomparable (Viitasalo 2005).

Another main principle was that the training and educating of soldiers should be based on researched knowledge. The respondents saw that so far the biggest reform of the Bologna process has been the emphasis on scientific scholarship. But what is the essence of being scientific? On the basis of what kinds of shared principles and ethos will the boundary between "civilian science", military science and the ordinary activities of the professional officers be negotiated and renegotiated by each individual researcher on a continuous basis? When the military researchers give reasons for e.g. the choices of textbooks (cf. choices of theories; various kinds of choices needed to be done during the research process), what kinds of reasons do they give, if any?

The researcher asked about the principles behind the choices of books, and the respondents expressed views like "the classics must be read", "centrality", "stimulating", "offering problems". Explanations related to scientific scholarship or some sort of "scientific ethos" were missing. Obviously much work needs to be done also on the meta-theoretical or philosophical level at the FNDC when the basic assumptions of social scientific research are scrutinized.

How about then the shared values of the personnel of the FNDC? During the Development Laboratory meetings the centrality of the values seemed to be a shared opinion, as expressed in excerpt 3: without values, enduring values, in a way without such energizing values this organization will not even sustain its vitality.

At the moment the FNDC claims that it values e.g. expertise. During the Development Laboratory meetings the researcher connected the ideas of shared enduring values, the espoused values of the FNDC (i.e. the expertise), "goodness" or "badness" of the teaching to the idea of an evaluation sheet for the teacher community (the Teacher's Self-Evaluation Sheet). The sheet allows the teachers to explicitly grasp their personal ideas about good pedagogical expertise and gives a practical example of how the organizational values could and should be operationalized.

During the thematic interviews the researcher tried to identify exactly and truthfully the present social situation at the FNDC and how the situation was seen from the managerial perspective. Each of the respondents was asked to depict *on paper* how he personally saw the present state at the FNDC (position A) and the desired future state (position B)?

The analyzed drawings showed misalignment of the interpretations of the present as well as of the future state of the FNDC. This point can be partly explained by the fact that at the same time the planning process for a new strategy was going on at the FNDC. Now after the new strategy has been completed, it will be interesting to focus on the means by which the identified misalignment (cf. e.g. Biggs 2003; Hakkarainen et al. 2004, 312) and basic assumptions of teacher Xs will be met in the complex system called the FNDC.

Analyzing the present situation at the FNDC

According to the main principles of the cultural-historical activity theory (CHAT), human behaviour is understandable only when interpreted against the background of the entire activity system (Engeström 1987, 2001; see Engeström 1987, 73-82 for the explanation how the structure of human activity has emerged through our evolution and cultural evolution). The activity system model offers us a useful alternative solution to conceptualize the social system for organizational developmental activities during e.g. Development Laboratory meetings (cf. the Change Laboratory Method; i.e. Engeström et al. 1996, 291; cf. in a school context Engeström et al. 2002), as it was done here.

In the first meeting no theories were introduced to the participants and the discussion was intended to be as practical as possible. The researcher emphasized that the aim was to develop the teaching practices by identifying some effective methods for societal guidance. The sheer possibility of this kind of task was severely challenged during the meetings and the following excerpt captures a widely shared opinion:

Excerpt 4

Especially in our project, but also in all kinds of activities aiming at developing learning, the basic problem and challenge is the question about the levels. Personally I could identify three possible levels: on the first level there are individual teachers, on the second level there is the system [the FNDC] and on the third level there is an outer level [the Finnish Defence Forces; or the outer area of operations]. The main problem is to choose the proper level, and complete unification of the three levels is not an easy task. Personally I prefer choosing one of these three levels to focus on.

In his comment the researcher explained his point of view by saying that the main question was not necessarily which one of the levels to choose but to identify the intertwined levels and act actively on several levels – the human behaviour is understandable only when interpreted against the background of the entire activity system, as will be shown below.

The researcher introduced the key principles (cf. e.g. Engeström 2001; cf. Mäkinen 2006 where also the social scientific position of the researcher has been explained) of the cultural-historical activity theory to the participants when the shared attention of the group had shifted towards the question about organizational levels. The researcher gave each participant a copy of the draft analysis of the FNDC made by him (figure 2).

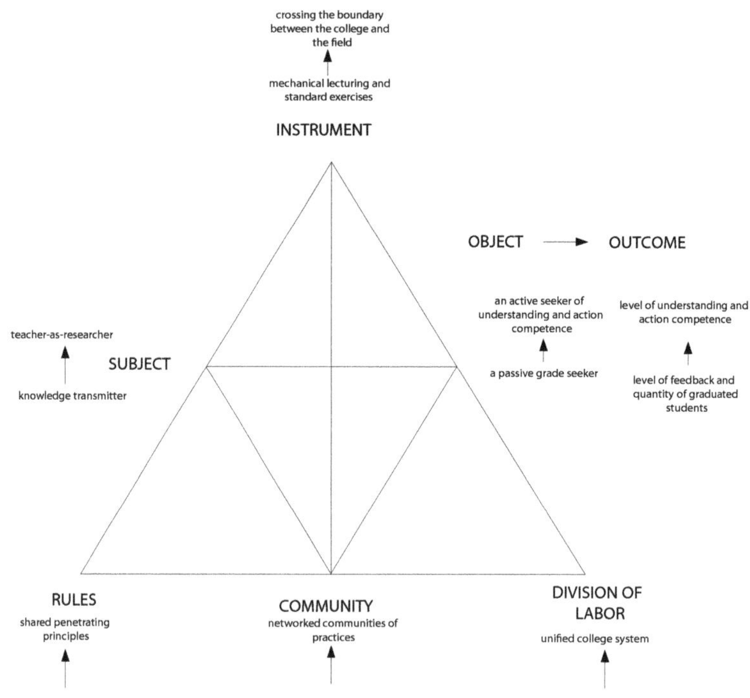

Figure 2. A draft analysis of the FNDC

The analysis shows how each corner of the activity system of the FNDC seems to need developmental actions to be *potentially* "lifted" to the qualitatively higher levels. But how well aligned with the basic assumptions and beliefs of the participating teachers was the analysis? Although the draft analysis was challenged by a simplified alternative (student → teacher → outcome), the gained experiences justify the initial interpretations of the researcher to a large extent.

Figure 3, an elaborated analysis of the FNDC, shows explicitly how well aligned with the interpretations of the participated teachers the draft analysis of the researcher actually was (cf. figure 3; next page).

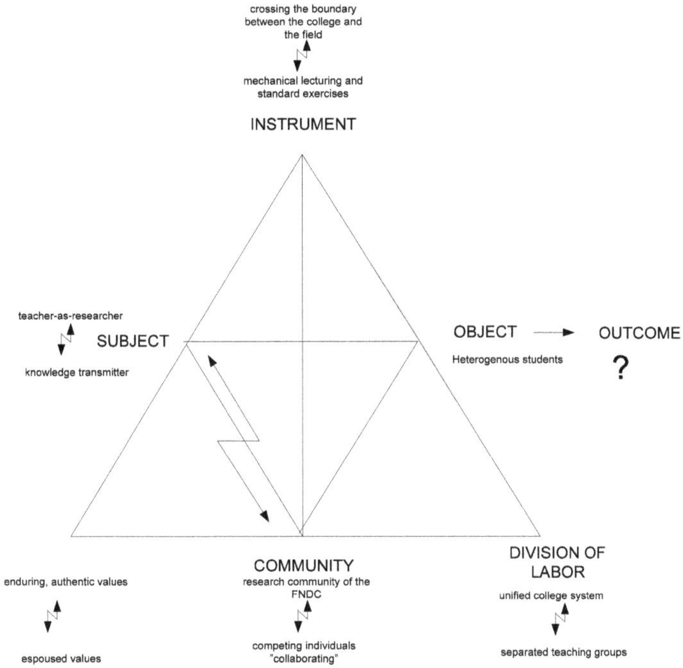

Figure 3. An elaborated analysis of the FNDC

There seemed to be some sort of agreement on the need states of the different aspects of the activity system of the FNDC, but despite of this, due to the latent secondary contradiction, the developmental mission was generally conceived as being impossible; at least for the teachers, but what about the managers of the FNDC?

The reasonability of the level of student feedback as a main result objective was severely challenged by the managers. Obviously the final outcome of schooling in the FDF should be something else than the "level of feedback" and it seems to be justifiable to ask to what an extent we can develop our military educational institutions on the basis of student feedback only.

In the discussion on the object of our schooling (i.e. the student), the managers said that the activeness and the responsibility of the students are highly valued. Recently the PISA (The Programme for International Student Assessment) reports have given some of the managers high expectations on the student population of the FNDC. Contrary to these kinds of expectations, the heterogeneity of the students was identified in a study made with the assistance of the Centre for Educational Assessment of the Faculty of Behavioural Sciences at the University of Helsinki in Finland.

The question of the increasing responsibility of the heterogeneous students is not an easy question at all, but despite its difficulty it needs to be faced on an ongoing basis by the teachers, departments, divisions and schools of the services and branches. The main idea of the need of a close look was put into an explicit form by colonel Douglas Macgregor when he analyzed the transformation of the Armed Forces of the United States of America: "It is unrealistic to expect that military leaders will demonstrate the requisite physical energy, mental agility, and moral courage in war to inspire subordinates to exercise initiative, to innovate, and to take risks if they have been discouraged from doing so throughout their military careers" (Macgregor 2003, 208). The question seems to be highly relevant also in the case of FDF and the FNDC especially at the time of increasing threats of international terrorism.

Presently the educational arrangements of the FNDC restrict the flourishing of the principles of activeness and individual responsibility. Putting the problems into the form of a question, we have to ask whether we will change our key principles or our educational practices in these cases.

Modelling our way out of the problematic situation

On the most simplifying level, development could be described as a movement from A (the current state) to B (the visionary final state) by Cs (by e.g. strategy). Naturally the effective use of even this formula demands that A is appropriately located "on the map".

For an educational institution, which is a complex social system, the transformational development means that *all* the teachers both individually and collectively self-reflect on the "goodness" or "badness" of their educational expertise. A Teachers' Self-Evaluation Sheet has been developed for these purposes (Mäkinen 2006).

After self-reflection, collective reflection is a fruitful next step along the long path of deep educational transformations. Double-hermeneutic (Giddens 1984, 1990, 2001, cf. also Habermas 1984) processes are the key issue for the following steps in teachers' on-the-job-education. A Military Teacher Education Course should be compulsory for "novice" teachers. Also the relative position of military pedagogy in the Senior Staff and General Staff Officer education needs to be reconsidered. The pedagogical expert status of the Department of Education[1] should be supported by all available means.

Life-long teacher education is a necessary but not sufficient prerequisite for the future aligned educational reforms and transformations (Mäkinen 2006). If one of the main educational principles of the FNDC is that the training and educating of soldiers is done on the basis of researched knowledge, it logically follows that every teacher should be a researcher, and more precisely, a progressive inquirer (cf. e.g. Dewey in Campbell 1995, Bereiter 2002, Hakkarainen et al. 2004, Hakkarainen et al. 2005, Hakkarainen and Palonen et al. 2004).

But a teacher is not just a researcher. He is also a leader. Therefore, the teachers' management and leadership development are essential parts of the teachers' on-the-job-education. The core expertise of the military teachers at the Finnish Military Educational System (FMES) *should* be pedagogically, managerially and also leadership oriented.

According to the espoused beliefs of the managers, the teachers play a central role in educational transformations. Now it seems to be the right time to turn these espoused expectations into systematic teachers' on-the-job education, especially when the alignment in research-based teaching seems to be the desired end at the FNDC.

When aiming towards future alignment at the educational institution, *figure 4* allows us to locate ourselves in the situation of the FNDC and its personnel.

Instead of being guided by the feedback of the military units and other profit centres of the FDF and the Finnish Border Guard (FBG), the FNDC should be aware of how powerful an impact their activities have on the key resources (i.e. knowledge) of their "military consumers".

Paradoxically, those practices for which people should be educated today *do not exist yet* (Hakkarainen and Palonen et al. 2004, 203). For the FNDC this means that it cannot solely focus on the present problems in the field (although it cannot neglect them either) but has to pay attention also to problems of the future. The FNDC is practically shaping the future of the FDF and the FBG, and the visionary future state of the FNDC has to guide the educational activities of the FNDC.

1 Since 1st of January 2009 the Department of Leadership and Military Pedagogy.

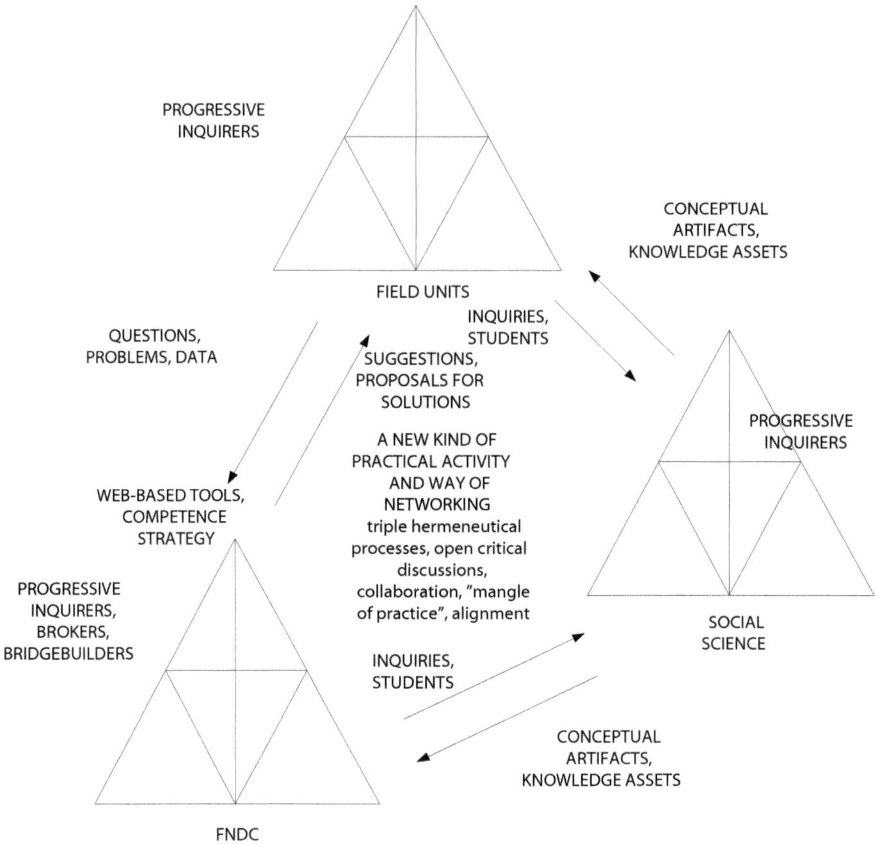

Figure 4. An example of the knowledge creating activity at the FNDC

The future of the FDF will be shaped by the individual actions of the progressive inquirers and their communities (i.e. communities of practices: CoPs; innovative knowledge communities: IKCs; see figure 1), and the main question is how to guide these kinds of activities. For these kinds of purposes the organization needs a vision – a shared knowledge vision (Senge 1990, Collins and Porras 1994, 219-228, Collins 2001, 108-110, Nonaka et al. 2001, Mäkinen 2006) made within the organization. Not just the content of the knowledge vision but also *the method how it is elaborated and what the ensuing actions of the knowledge vision are*, are crucial and fundamental questions for the FNDC.

The FNDC has added the concept of "human security" and "learning organization" to its strategy (to the "knowledge vision"), presumably being followed by collective progressive inquiries into the phenomenon determining how the knowledge base of the FNDC and the FDF will be evolving in the long term. It seems to be obvious that human security will be guaranteed by the collective effort of several administrative branches and non-governmental organizations. But what this kind of knowledge vision means to the FNDC and for the education of officers, needs to be elaborated, because it will not emerge out of the blue. For the FNDC this requirement means that it has to *justifiably extend its networks* to be used for knowledge creating activities and by its progressive inquirers.

Examining and implementing the model in practice

The Teachers' Self-Evaluation Sheet has been developed since 2001 and during the past years the Department of Education of the FNDC has had practical experiences of its use. The practical meaning of the sheet will naturally vary when it is used, but so far it has been used as an additional tool in annual planning discussions at the FNDC. The deepest layers of the educational institution are not neglected anymore, and the harsh social reality of the teachers is faced and improved.

The systematic use of the evaluation sheet allows us individually but also collectively to reflect on the "goodness" and "badness" of our teaching expertise and to articulate tacit knowledge into explicit knowledge (Nonaka and Takeuchi 1995, 62-70; Nonaka and Toyama 2003, 495-498). Further on, its use will allow us to increase our interdepartmental and inter-organizational connectivity, which seems to be a crucial success factor for an educational institution in the Knowledge Age.

Ultimately this kind of knowledge creation process allows us to reflect on our deep-seated cultural basic assumptions about the "machine-like bureaucracies" claimed to be "static" instead of "dynamically organic". Instead of being fully hierarchically controlled in a top-down manner where the decentralized nature of complex social systems is neglected, the shared responsibility of the teachers and the other personnel working at the FNDC will be hopefully emphasized in the future.

References

Allee, V. (2003). *The future of knowledge: increasing prosperity through value network.* Amsterdam: Butterworth-Heinemann.

Anderson, P. (1999). "Complexity Theory and Organization Science". *Organization Science,* Vol. 10, No. 3, 216-232.

Bereiter, C. (2002). *Education and Mind in the Knowledge Age.* New Jersey: Lawrence Erlbaum associates.

Biggs, J. (2003). *Teaching for Quality Learning at University. Second Edition.* UK: The Society for Research into Higher Education & Open University Press.

Bologna Process, [On-line]. Available: http://europa.eu.int/comm/education/policies/educ/bologna/bologna_en.html.

Boudon, R. (1982). *The Unintended Consequences of Social Action.* London: Macmillan.

Campbell, J. (1995). *Understanding John Dewey: Nature and Cooperative Intelligence.* US: Open Court Publishing Company.

Collins, J.C. (2001). *Good to great: Why Some Companies Make the Leap and Others Don't.* New York: HarperBusiness.

Collins, J.C. & Porras, J.I. (1994). *Built to Last: Successful Habits of Visionary Companies.* UK: Random House.

Engeström, Y. (1987). *Learning by Expanding: an activity-theoretical approach to developmental research.* Helsinki: Orienta-Konsultit.

Engeström, Y. (2001). Expansive Learning at Work: toward an activity theoretical reconceptualization. *Journal of Education and Work,* (14) 1, 133-156.

Engeström, Y., Engeström, R. & Suntio, A. (2002). Can a School Community Learn to Master Its Own Future? An _Activity-Theoretical Study of Expansive Learning Among Middle School Teachers. In G. Wells & G., Claxton (eds) *Learning for life in the 21st Century.* UK: Blackwell Publishers.

Engeström, Y., Virkkunen, J., Helle, M., Pihlaja, J., & Poikela, R. (1996). Change laboratory as a tool for transforming work, In G. Ruckriem, (ed), *Developmental Work Research: Expanding Activity Theory In Practice.* Berlin: Lehmanns Media.

Giddens, A. (1979). *Central Problems in Social Theory: Action, structure and contradiction in social analysis.* Hong Kong: Macmillan Press.

Giddens, A. (1984). *The Constitution of Society. Outline of the Theory of Structuration.* Berkeley, Los Angeles: University of California Press.

Giddens, A. (1990). *The Consequences of Modernity.* Cambridge, UK: Polity Press.

Giddens, A. (2001). *Sociology. Fourth Edition.* Milan: Polity Press.

Habermas, J. (1984). *The Theory of Communicative Action: Reason and the Rationalization of Society.* Boston: Beacon Press.

Hakkarainen, K., Bollström-Huttunen, M., Pyysalo, R. & Lonka, K. (2005). *Tutkiva oppiminen käytännössä: Matkaopas opettajille. (Progressive inquiry in practice: A guidebook for the teachers).* Porvoo: WS Bookwell (in Finnish).

Hakkarainen, K., Lonka, K. & Lipponen, L. (2004). *Tutkiva oppiminen: Järki, tunteet ja kulttuuri oppimisen sytyttäjinä.* (Progressive inquiry: reason, emotions and culture as inspirers of learning). Porvoo: WS Bookwell (in Finnish).

Hakkarainen, K., Palonen, T., Paavola, S. & Lehtinen, E. (2004). *Communities of Expertise: Professional and Educational Perspectives.* Amsterdam: Elsevier, Earli.

Jackson, M.C. (2000). *Systems Approaches to Management.* New York: Kluver Academic / Plenum Publishers.

Kupers, R, (2001). What Organizational Leaders Should Know about the New Science of Complexity. *Complexity,* (6) 1, 14-19.

Lindblom-Ylänne, S. & Nevgi, A. (2003). *Yliopisto- ja korkeakouluopettajan käsikirja. (Handbook of University Teachers).* Vantaa: WSOY (in Finnish).

Macgregor, D.A. (2003). *Transformation Under Fire: Revolutionizing How America Fights.* Westport, Connecticut: Praeger Publishers.

Mäkinen, J. (2006). *The Learning and Knowledge Creating School: Case of the Finnish National Defence College.* Helsinki: Edita Prima.

Monge, P.R, Contractor, N.S. (2003). *Theories of Communication Networks.* New York: Oxford University Press.

Nonaka, I. & Takeuchi, H. (1995). *The knowledge creating company.* Oxford: Oxford University Press.

Nonaka, I. & Toyama, R. (2003). The knowledge-creating theory revisited: knowledge creation as a synthesizing process. *Knowledge Management Research & Practice,* Vol. 1, 2-10.

Nonaka, I., Toyama, R. & Byosiere, P. (2001). A Theory of Organizational Knowledge Creation: Understanding the Dynamic Process of Creating Knowledge. In M. Dierkes, A. Bethoin Antal, J. Child & Nonaka, I., *Handbook of Organizational Learning & Knowledge.* UK: Oxford University Press.

Sahrouris, E (2000). *Earthdance: Living Systems in Evolution.* US: University Press.

Senge, P (1990). *The fifth Discipline: Art and Practice of the Learning Organization.* New York: Currency Doupleday.

Viitasalo, M. (2005). Tutkintojärjestelmäasetus on muuttunut. (The Changing Government Decree on University Decrees). *Sotilasaikakauslehti,* 80 (8), 56-57 (in Finnish).

Wenger, E. (1998). *Communities of Practice: Learning, Meaning and Identity.* Cambridge: Cambridge University Press.

Wheatley, M. (1999). *Leadership and the New Science: Discovering Order in a Chaotic World.* Second Edition, San Francisco: Berrett-Koehler Publishers.

Wolfgang Royl

On the Mutual Compatibility of Obedience of Orders, Comradeship and Conscience

The term "interoperability" is taken beyond its purely technical meaning to be applied to the field of military pedagogy. The question is raised whether male and female officers learn how to actively deal with the tension of seeing a subordinate as a comrade on the one hand, while having to give him orders which may expose him to the risk of being wounded or even killed on the other. A similar situation of conflict may arise in peacetime if an order is given which a soldier cannot reconcile with his conscience. If he refuses to obey, he might risk his career. What then may a military superior demand from himself and from his subordinates?

Military training which aims at dealing with such dilemmas effectively and in a learning goal-specific way will benefit from studying both the findings of subject-matter experts on comparable relationships of trust and the lessons to be learned from their expertise. The soldier's profession requires him to take an oath to risk life and limb for his nation, which makes a soldier's relations with his superiors and the defence mission as a whole particularly sensitive. The moral dimension of this relationship also concerns the military top leadership, as was vividly shown by the war crime court proceedings in The Hague.

In terms of military pedagogy, constellations reflecting the triangle of comradeship, obedience of orders, and conflict of conscience become worthy of discussion. It is not a coincidence that the German *Bundeswehr* pursues a military doctrine referred to as *Innere Führung* (Leadership and Civic Education) designed to accomplish the challenge of exerting a deliberate influence on the personality development of military leaders both during their training and during continuation training phases. The conflicts that may nevertheless arise were amply demonstrated by the court case, which the *Bundeswehr* instituted against Major Z., who could not take it upon his conscience to develop software that might be used in the Iraq War. He had refused to obey an order to that effect. This is an exorbitant example, which reflects the mental interoperability between the parties to the court case – the German Ministry of Defence and a military officer – in the 126-page decision taken by the Federal Administrative Court on 21 June 2005. The officer was acquitted and his reduction in rank rescinded.

Here, interoperable behaviour is related to thought processes in which the soldier and the superior are comrades on the one hand, but the military mission demands obedience of orders on the other. A military superior will expect obedience, both as a matter of principle and because he depends on orders himself. If a subordinate has the right to refuse an order when he has justified doubts as to the rightfulness of a military mission, what will be the consequences for military leaders, military training and the operational readiness of the troops?

The Ambiguity Dilemma

Being a Military Superior and Comrade Both at the Same Time

As soon as a behaviour situation is determined by antagonistic value orientations, moral dilemmas are bound to occur. Whereas comrades will help each other to survive in a dangerous situation, a military superior must be prepared to even give an order that may require the soldier who has to obey it to risk his life. In such a case, both the individual soldier and the team are confronted with the fact that they must take on even a deadly mission if the combat situation so requires, knowing full well that being a soldier is not exactly a life-insurance policy, anyway. Nevertheless, a lot has changed since 54 BC, when the salute by the Roman legionaries was *Caesare, morituri te salutant*!

It is known from war literature that in the past an officer was required to shoot on the spot any soldier who refused to execute a dangerous mission. This was to prevent the fighting morale of the unit concerned from being weakened by somebody refusing to obey orders. Only afterwards, during military tribunal proceedings – if they did take place at all – was it possible to judge on the lawfulness of such a shooting. For those fighting directly at the front, there was not even time for court-martialling them on the spot. The French General Duchêne is said to have given the order to pick out and shoot mutinous soldiers right there and then in July 1917.

In the army of Frederick the Great, there was the maxim that in a combat situation the grenadiers should fear their own officers, who followed the advancing lines, more than they feared the enemy so as to keep them from even thinking about seeking refuge in flight.

Military history tells us that this way of regulating the mortal danger of the soldiers only changed with the development of conscript armies, whose fighting morale, so we are told, was characterised by the will to defeat the enemy. They confronted danger from inner motivation. Such a *levée en masse* has not repeated itself. Reports about the Red Army reveal that during World War II the Soviet soldiers, who often had been recruited against their will, were closely monitored by their officers, who would follow them right behind the battle line during an assault operation and were under orders to shoot anybody trying to leave the danger zone without permission.

The current situation in military operations is such that even small numbers of soldiers killed in action may cause the public at home to withdraw their support of the military campaign. This shows the degree of political sensitivity that has developed with respect to friendly casualties in military operations. Today, the democratic public is very much aware of every single dead soldier. Consequently, the tactical design of UN missions focuses on avoiding friendly losses altogether, if possible, while even accepting the risk of achieving the officially declared military objective of the operation only to a limited extent.

Fortunately, at the moment there is peace in NATO's area of responsibility. The present politico-military situation excludes the onset of a war there, which the military leadership would expect to entail considerable losses of human lives and materiel. Also from the point of view of military subordinates, no threat situation can be imagined today which might lead to military combat situations like the ones millions of soldiers commonly experienced on all fronts during the two world wars of the past century.

However, also in peacetime there are moral dilemmas that may serve to increase the awareness of the significance of ethics in military actions. They should not be taken any less seriously than the ones occurring during a combat mission. In the global spectrum of operations of armies, and even when troops alternate between the area of deployment and their home garrison every six months, the mental transition between the phases of pre-deployment and operations is a gradual one. Whether a military mission is morally justified, and also what consequences it might have for oneself as a person, are questions some soldiers ask themselves even well before they receive their travel orders. These soldiers, even if they are of relatively low rank, are sometimes in a position to develop political opinion leadership simply because the group dynamics among their fellow soldiers work that way.

The officer, and also the informal leader of a group, normally differ from other soldiers in that they have a larger overview of military everyday routine and, due to their superordinate interests, also develop an ability to judge the political context of military matters. Those who are, or seek to be, considered as opinion leaders among the enlisted ranks or among their fellow officers must accept responsibility for the influence they exert on others by the views they express. The relationship between the official announcements made by the military leadership and the unofficial comments made by the addressees of such messages affect the fighting morale and readiness for action. The difficulty and significance of military leadership increase in accordance with the spectrum of such dialectics. Experience shows that this problem partly consists in the fact that leaders at the supreme military command level cannot easily assess how their decisions are communicated to the troops, and how well they are accepted by them. However, any carefully drafted order of the day may improve the soldiers' mental state, and thus the acceptance of the mission, provided that it is duly noted by those it is intended for.

Hierarchy of Dilemmas

Just like elsewhere, there are simple and complex conflicts in military everyday life. The simple ones are those that arise if a soldier wants to unfairly force his will on others in a section or troop. This may involve verbal attacks, or he may employ his physical superiority. A person may also get into difficulties if somebody denies him information. Since, as a rule, there is a high contact density among the privates sharing a single room, it is of major importance whether somebody feels treated unfairly by his comrades or his direct superior. In such cases, it is the responsibility of the senior soldier of the room and the NCO to balance the psychosocial situation, thus restoring stability of group cohesion. In the *Bundeswehr*, such conflict situations may have to be resolved by the Parliamentary Commissioner for the Armed Forces, should they persist. In today's Russia, the situation is still different. There are reports about older soldiers establishing what is referred to as "the rule of the grandfathers" (dedovshchina); sometimes terrorising recruits so badly that the victims choose suicide over continuous torture. By way of political protest against these intolerable conditions, the "Soldiers' Mothers" movement was created, which investigates such cases of subjugation and even reaches the Russian Ministry of Defence by informing the public. Orders from higher command to stop these harassments and to call criminal groups to account obviously have had no effect so far because the lower ranks in charge of recruit training profit from the fact that their work, i.e. to condition sensitive recruits for the hardships of military missions, is done by others.

The higher the leadership position in the armed forces, the more complex the dilemma situations become. In special cases where, for political reasons, a minister of defence does not see another solution to the problem, a general – or even two, for that matter – who has come under public fire, may, for example, be made to retire early. This is how in the German *Bundeswehr* lieutenant generals Hans-Heinrich D. and Jürgen R. had to face compulsory retirement. As could be read in an article by the *Frankfurter Allgemeine Zeitung* (FAZ) of 26 January 2006, the two generals had reached an advance agreement in the disciplinary proceedings against the son of one of them, which had just started. As fellow soldiers had reported, the general's son had caused a stir with political statements that contradicted the official views of the military leadership. This way of behaving politically incorrect as an officer was made known to journalists, who made it public.

This example may serve to illustrate the moral sensitivity that has to be kept in mind also among the officers' corps. The explosive nature of moral dilemmas increases with the ranks of the military personnel involved. It is caused not only by offensive political statements as such but also by their uncontrollable effects inside and outside the armed forces. The ordinary citizen may ask himself what might be the prevalent cultures of political opinion-forming in generals' circles, and how this influences their families and sons, who, already at the beginning of their career in the army, firmly expect to climb the ladder of success. The alternative of solving such a problem case by measures of *Innere Führung* is more or less impossible for a moral dilemma on this level of hierarchy. The German Minister of Defence was thus well advised to do as he did. He had to discharge two of his generals despite the fact that they had comprehensive technical competence and had served the Army in a responsible manner up to then. Afterwards, his duty-related and moral decision was additionally justified by the fact that the two generals concerned went to court and lodged an objection against the minister's decision. It's no use to sue for victory once the battle has been lost!

No sanctions will probably be imposed against those superiors who did not take timely and adequate action against the personal presumption of the general's son for fear of making themselves unpopular with superiors further up the chain of command. In this respect, this case also casts a light on the moral quality of military everyday culture, which obviously supposes that one's own behaviour should be in keeping with what higher military superiors might expect, and that this a part of professional ethos even if they evidently act selfishly.

Such a conflict is – and this must be considered, too – a lesson teaching us about the moral functionality of free society. Even if the journalist may have attacked the military merely from base moral motives, he did contribute to giving morality and law a chance to take effect, much as if he had been led by the "invisible hand" according to Adam Smith (1776).

The Responsibility of Decision Making

A historic example of a military decision that was made for the common good is General von Choltitz' failure to obey instructions towards the end of World War II. The general had not executed Hitler's order to defend Paris against the advancing troops of the Allies. According to the German law of war of the time, the general had thus forfeited both his rank and life. As can be seen from the general's memoirs, the decision to disobey an order given by the *Führer* had not been easy for him. By surrendering Paris to the Allies without a fight, von Choltitz gave priority not only to the survival of his own troops but also to the protection of a world cultural heritage. He did this although his rank obliged him to obey orders given by his superiors. Hitler's personal order had been to reduce Paris to rubble.

In those last days of the war, there also were cases of military commanders ordering the shooting of hostages although it could not be expected that this would decisively influence the outcome of the war. The fact that people who were responsible for making the decision to order executions by shooting were convicted in war-crime trials does not only cast a light on the inhumane actions carried out but also on the decision-making conflict. People who were not involved in partisan actions but were encountered in the area were killed as an act of deterrence in order to discourage the partisans from further attacks on the fighting units. Today, retaliatory attacks by the Israeli Air Force assume this function with respect to Hisbollah fighters.

In retrospect, it is difficult to judge whether, from the point of view of the decision-makers, such decisions were indeed just, or only in line with standard procedures. This also applies to the irregulars, the resistance fighters themselves, who could not have failed to notice that their raids were followed by immediate acts of retaliation against the civilian population. More often than not, the primary victims of the reprisals had not been involved in the conflict at all.

Excursus: Does the training, which military leaders receive with respect to the execution of a mission and/or an order, differ from the training undergone by executives in a major industrial enterprise? Not really. Both professions require the ability to make decisions – in the superior interest of the armed forces or of the economic situation of a company – which, from the point of view of those concerned, may have profound effects. Taking measures such as passing over a fellow soldier in a promotion or dismissing an employee must not break the heart of the superior who has to carry them out. Therefore, the way he learns to shield himself against feelings of sympathy will, as a rule, be reflected in his body language: Successful superiors must be trained to appear unapproachable, both in their body attitude and the language they use in giving directions. Keep the chin up, maintain a focused gaze, and move your hands a little abruptly when unpopular decisions need to be communicated. Authority can be learned, especially from the example of colleagues who have already climbed the career ladder.

However, the military has a special position compared with a private enterprise. A particularly high degree of confidence between superiors and subordinates is required there. Unlike in the civilian sector, the soldier's tool is the weapon. It may be employed not only against the enemy but also against an unpleasant superior who has turned into a personal enemy. It is known that in the Vietnam War, military leaders who did not behave like comrades but brutally played the superior had to fear being killed by their own people. “Fragging” is the special term used for shooting a military superior from behind. The phenomenon itself, i.e. killing hated officers of one’s own side, is reported also from World War I where Poles from Western Prussia and the province of Posen, who had been recruited forcefully, were committed at the front in France under the command of German officers.

This experience from war is a warning example of demanding from subordinates to obey orders without the existence of a relationship of comradeship between them and their superior. However, a practice that promotes cohesion is looking for missing soldiers and getting wounded comrades out of the danger zone, although such rescue missions may cause additional casualties. Part of each soldier’s mental interoperability, however, was the certainty that his comrades would not let him down in an emergency situation.

Putting aside the fear about one’s own survival in order to search for a missing comrade and recover wounded fellow soldiers acquires personal interest in so far as each member of a combat team could encounter the same fate. In this case, interoperability develops in a two-fold way. The soldier overcomes his personal desire to rather avoid the additional mortal danger involved in helping others and integrates it into a mutual moral relationship. It corresponds to what the Golden Rule of Morality says: “Do unto others as you would have them do unto you.”

In an emergency situation where somebody’s life is jeopardised while trying to save another person, this process of putting one’s own life first and then thinking of the other person remains an ongoing one. It seems to make sense to call it “mental interoperability”, since it is advisable to refer to possible dilemma situations and associated processes already in military education. The cognitive information is thus transferred to layers of consciousness enabling the student to imagine his behaviour in the described reciprocal relationship between his own desire to survive and the possible sacrificing of his own life for a fellow soldier. This means putting the social relationships in a military unit to the test even without a real danger situation: Whom would I stand by even if this meant putting myself into danger?

Giving and Taking Orders

§11 of the Legal Status of Military Personnel Act obliges the soldier to show obedience. One is particularly reminded of this rule if a soldier refuses to execute an order. In such a case, he violates a law and must be held responsible.

How does such a relationship based on the obedience of orders influence thought processes? Is an order executed without complaint, or even gladly? Does an order trigger inner resistance? Is this resistance directed against the person giving the order or against the content of the order? Does the superior always know how to adequately communicate the order he has just received himself to his subordinates? For this case, standard routines are common in military everyday life. E.g., it is an established procedure to repeat an order so that the person giving the instruction can check whether he has been understood correctly.

Personal charisma has a direct positive effect on leadership. The relationship between the military leader and the soldiers led may be such that they would go "through fire and water" for him. But the majority of military leaders – who have not been born with leadership qualities but had to *learn* how to lead people in a military environment – still can address their subordinates personally and open up individual perspectives for them which may occur within a military career or already in civilian life, over which the officer has a better overview due to his academic background. In addition, there is what military pedagogy calls working on a person's personal development. For recruits who, in their private lives, have not really succeeded in fighting down their weakness of will, the officer's orders make up for their lack of ego strength. He uses orders so to speak as a means to introject into the recruit's ego his own ego strength. In executing the orders given, the recruit experiences how it feels to carry a mission through to completion and to develop stamina in routine work such as cleaning weapons, keeping his locker tidy, polishing his shoes until they shine and keeping his uniform in order. Mothers whose sons lacked tidiness are said to have warned them: "You just wait until you join the army. They will teach you there!" Such popular belief sometimes reflects expectations that military service will make up for educational aims that could not be achieved in the family. But this is not the true task of military service. However, similar as in school, the military sometimes also must teach its recruits basic forms of behaviour they failed to acquire. This way, the military leader may become a kind of auxiliary ego for the recruit, whose ego expansion is initiated through military general education. Due to the UN missions, the latter also comprises a political view of globalisation and the associated co-responsibility for the military safeguarding of world peace. As a result, officers today must accept a more comprehensive co-responsibility for education (education-oriented leadership) than in the past, a task that must be performed in addition to the special military training and education that has to be provided. Accordingly, it may happen that a soldier benefits from a good superior also in his personality development and thus for his whole life.

Command Authority

A superior who gives an order to a subordinate expands the range of his own options of duty-related action to include those actions he has ordered and thus obliged his subordinates to do. In the interest of the military mission and according to his rank, the military leader will mass the combat power of a troop, a battalion or a division in order to reach a target. The higher the command authority, the more it is executed from the rear headquarters of higher command. The lower ranks, however, lead from the front.

Due to the hierarchy of command authority up to the supreme command of an army, the military leaders bearing the main responsibility for the implementation of strategic plans are spared from having to look directly at the things happening on the battlefield. As far as war movies can illustrate the atmosphere of a military headquarters or large-scale exercises can give a realistic impression, the soldiers there are not "moving under fire" but, as staff officers, to and fro between maps on which troop movements are followed and marked with symbols. Unlike in a live experience of a combat situation, military leaders on higher command echelons work with symbols as well as with strategic and tactical images of combat action, which relieves them from having to experience horror, the noise of the battle and fear of death. This is an important prerequisite for focusing one's thoughts if existing military plans must be co-ordinated with the respective combat situation as reported from the theatre of operations. The chain of command, which modern means of communication have perfected by now, must in addition guarantee continuous, fast communications between the command centre and the soldiers in the field. As the significance of such command machinery and the command authority associated with it grows, the importance of the supreme commander increases, too. Visible signs that serve to make both fellow military personnel and civilians aware of this are the rank insignia as well as the number and origin of military decorations somebody has been awarded. Historical paintings of generals and field marshals in resplendent parade uniforms and on noble horses getting ready for a virtual jump not only serve to illustrate the military's grandeur. They are always also meant to remind us of the responsibility for the troops commanded, who risked life and limb to execute the orders of these military superiors. Military awards and decorations are outward signs of leadership responsibility.

Disobedience to Orders

Military superiors, who themselves receive orders and execute them, take it for granted that their subordinates will do the same. In this sense, the military's functionality is based on the principle of command and obedience, which has proved its worth. This military principle is not emphasised on every occasion in military everyday life. Many things soldiers and their superiors have to do are just natural routine processes in a military environment. From the wake-up call until the end of duty, the duty schedule – which may be very detailed – determines what an individual company has to do throughout the day. The soldiers know how to act in case of alert exercises. Orders of the day are issued to major military units to regulate their interaction.

In cases where individual soldiers of the *Bundeswehr* intentionally disobey orders, the consequences stipulated by the German Military Penal Code are applied. It is applied by disciplinary courts and disciplinary lawyers. This will happen e.g. if somebody is absent without leave, uses military materiel for another than the intended purpose or becomes violent towards fellow soldiers and superiors. §20 of the Military Penal Code explicitly refers to "disobedience to orders":

> "A prison sentence of up two three years is imposed upon anybody
> 1. who refuses to obey an order by rejecting it in word or deed, or
> 2. who insists on disobeying an order even though the latter has been repeated."

In an army like the German *Bundeswehr* where, due to the principle of *Auftragstaktik* (mission-type command and control), not only military leaders but also their subordinates are expected to think for themselves and form an opinion of their own, it comes as no surprise that critical ability with respect to decisions by one's own military superiors is increasing, too. It is the task of the Parliamentary Commissioner for the Armed Forces to investigate complaints by soldiers before the Military Penal Code is applied. In addition, such complaints help identify measures of *Innere Führung* that are in need of improvement, or they may reveal that there simply was a misunderstanding of local command decisions.

The fact that in an army, cases of complaints and even crimes occur is no reason for concern as long as they are within certain limits. The French sociologist Emile Durkheim (1858-1917) left us an important investigation on what he referred to as "deviant behaviour" in a society. He states that societies will function best if their statistical base value of deviant behaviour is neither lower nor higher than a certain value. If laws were such that they absolutely prevented criminal behaviour, they would paralyse the citizens' spirit of entrepreneurship because they would deny them any leeway to take action. Invariably, deviant behaviour of citizens also provides an indication as to which social regulating mechanisms need to be examined with regard to their consequences. In the following, a case of disobedience to orders in the German *Bundeswehr* is described to illustrate the far-reaching consequences the deviant behaviour of an officer had.

Interoperability of Character Elements

Obedience, comradeship and conscience are combined in a unity of personality in an exorbitant example. This is done with the intention to describe and explain the special character of the military. For this purpose, a judgement passed by the supreme German administrative court is used as an example, whose judges decided in favour of a soldier who stood by his own conviction although this meant taking a high personal risk. The legal proceedings against him are also an example of how the separation of powers takes practical effect in a democratic state.

Triangulation

The relationship between obedience and comradeship is illustrated using a case here, which is both an example of how a constitutional state executes control over its army and acknowledges decisions a soldier has taken based on his conscience as positive. The behaviour elements of obedience, comradeship and conscience are triangulated in preparation of the case analysis. This combination of concepts represents an individual decision context, which is to be used in order to reconstruct how a person stabilises his behaviour consistency, his character in a difficult conflict situation. For this purpose, the triangulation method described by Flick (2004) is complemented by a hermeneutic variant. As a result, the combination of concepts becomes a combination of meanings, which is of particular relevance to the character of the military. The application of this triangulation approach yields three connections (see fig. 1) between three behaviour elements, which are to be used to reconstruct courses of thought reflecting the character of the soldier lodging the complaint.

Making obedience, comradeship and conscience the corner points of the triangle needs to be explained. What correlation can be experienced between obedience and comradeship, what interoperability is there between those two aspects? In the present case, this question is related to the military field of relationships, i.e. a military superior who gives an unpleasant order to a subordinate. As a rule, such an order is based on situational requirements. At the same time, however, the order also concerns the relationship of comradeship between the person giving the order and the one who is to execute it. Will the soldier receiving the order consider it justified that he and not somebody else was chosen for the delicate military mission necessitated by the situation? By giving this order to him, does the superior perhaps spare another soldier who would in fact be equally qualified to take on the difficult mission? If, as in this case, the interrelationship between obedience and comradeship is concerned, the conscience function plays a role as an intervening variable. Does the soldier concerned feel treated fairly because this time it is his turn to execute a difficult mission similar to the tasks other soldiers had to accomplish before him? Does he trust in his superior as he would in a comrade whom he can rely on not to intentionally put him at a disadvantage when allocating difficult missions? As a rule, subordinates sense by intuition whether they are treated fairly, and so do military superiors with respect to those officers who may give them orders. Whereas the subordinate judges the conscience of his superior through his need to be treated fairly by the former, the superior in turn must be able to rely on his soldiers not to double-cross him. The military superior, too, has a right to be treated fairly by his subordinates. They are not supposed to exploit possible leadership weaknesses. They must not let him down in executing what his own superiors have ordered him. By giving an order, the military superior also appeals to the comradeship of the person having to execute it.

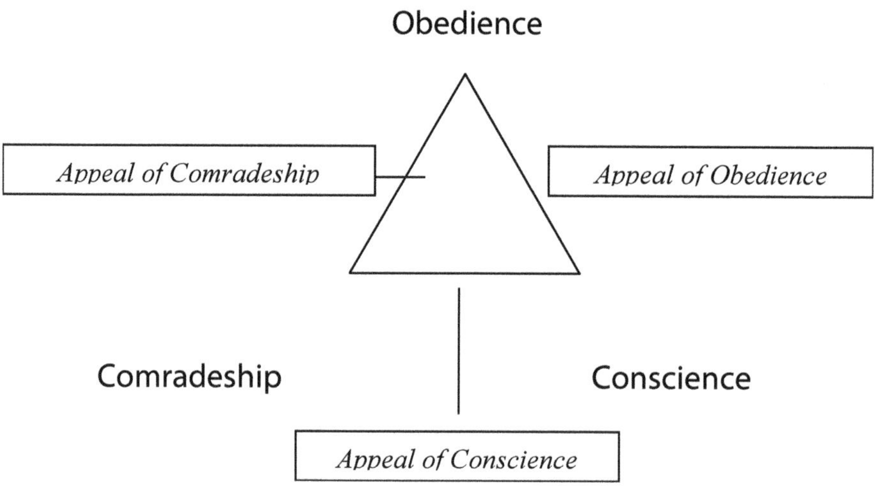

Figure 1. Elements of Character, triangulated

It is claimed here that the interoperability between comradeship and obedience is controlled by two authorities of conscience, i.e. of the superior and of the subordinate. The fact that this obviously really is the case leads to the military pedagogic task to explicitly appeal to the conscience function of both superior and subordinate, and to develop and foster it. By doing so, the military pedagogue so to speak implements Kant's moral law in the military. The corner points of obedience, comradeship and conscience are linked to form a triangle by the triple appeal: Be obedient! Be a good comrade! Be fair!

Such an interoperable character constellation would be taken for granted and thus be considered not requiring an explanation if the reality of working and living together in the military fully corresponded to this situation. However, the communicative network of relationships in military everyday life, in which mental correlations frequently act without talking about them, always also provides superiors and subordinates with opportunities to circumvent ethical basic attitudes such as obedience, comradeship and conscience in their practical behaviour. There may be various causes for evading moral responsibility for the stability of ethical basic attitudes. In any case, negative experience in the way subordinates and superiors treat each other loosens the cohesion among the troops and weakens the persuasive power of military education and training.

An Example of Application

A major, who was a specialist for software development, had refused to assume a responsible position in developing a highly complex computer program that was to serve the "optimisation of maintenance, repair, supply, transport, data processing, information, command and control, situation assessment and accounting". It had become known at the military installation that such a program might be used, among other things, for the logistic support of the western units in the Iraq War. Since this 2nd Iraq War was not conducted under a UN mandate but by the U.S.A., the United Kingdom and collaborating nations out of their own initiative, the major had refused to co-operate in this IT mission. He objected to participating, through his work and even if only indirectly, in acts of warfare that were against international law. Due to this refusal of obedience, the competent *Bundeswehr* disciplinary and complaints court demoted the major to the rank of captain. He had appealed against this judgement, and the lower court's decision had been reversed by the Federal Administrative Court in Leipzig after having passed through the proper instances. The loss of rank was rescinded by the decision of 21 June 2005. The supreme administrative court of the Federal Republic of Germany found for the plaintiff.

It is not at all true that this judgement by the supreme German administrative court now will cause large numbers of soldiers to express their own view on the Iraq War by starting to critically look at duty-related tasks. The rule will rather be, just like probably in all armies, that soldiers hold the view that responsibility for orders rests with the superiors who issue them, who in turn trust their higher commanders, with the Federal Minister of Defence at the top level of command. Independent thinking is, in fact, appreciated by the armed forces. However, it must not interfere with the respective duty routines. Since, in democratic states, armies are subject to parliamentary control, it must be ensured that law and order are observed, anyway (Royl 2007). This is something the parliamentary opposition has a particular eye on. For conflicts in military leadership, the Federal Republic of Germany in addition has the Parliamentary Commissioner for the Armed Forces, who is directly appointed by the German *Bundestag*. He is not bound by instructions.

What makes the case of Major Z. special is the fact that he felt himself to be in a position to judge whether the things he did were in line with the international law of war and the UN Charter. At first glance, an outsider may think the major to be an officer of the hair-splitting type or somebody who likes to shirk difficult work. Such doubts, however, are not justified. The major had received good performance ratings in various duty assignments. Without any problems, he had received a commission as a career officer already years before.

However, how come that this officer inconveniences himself and his superiors? Or, in other words: Why are there actually not more soldiers who, if necessary, are ready to criticise what they are ordered to do?

In this case, a personality analysis should be conducted, which seems to be what actually happened because after his refusal of obedience the major was first sent to see a psychiatrist. In the presentation of facts by the federal administrative court, it can be read that the major had to take off all his clothes although there were no physical complaints to be examined. If this undressing was a measure meant to render the soldier defenceless and to unsettle him, it failed. Psychological tests carried out in order to find out about psychopathological peculiarities did not indicate a pathological result, either. The man proved to be in good psychological and physical health. There was no reason for punishment because, as stated in the court's decision, the refusal to provide a logistic service in support of a war violating international law had been justified. In fact, the consequence would have to be that all the others who did not refuse to collaborate in the software program have made themselves liable to prosecution. This, however, was not the subject of the proceedings, which only dealt with the question whether the officer's demotion had been right or wrong.

From the point of view of military pedagogy, the more important question is how such a character development takes place, and what value such a soldier has both for the armed forces and society. It should also be considered what action should be taken as a consequence of the case analysis, and if there are even lessons to be learned.

One clue is given by the major himself. In the court's decision, it is mentioned that during the time of the disciplinary proceedings, he appeared for duty with a white cloth flower and in civilian clothes. By this, the major obviously wanted to show his affinity with Sophie and Hans Scholl, who had fought against the Nazi regime by distributing leaflets at the Munich University and using the symbol of a white rose, for which they were sentenced to death in 1943. Comparing oneself with the victims of the Nazi rule would be considered completely inadmissible by an objective observer. The federal judges, however, showed subtlety in acknowledging the major's subjective gesture by mentioning his internal demonstration with the "white flower" on page 17 of their decision. The major uses this symbol to make a stand against a regime, in this case the Federal Republic of Germany, that in his opinion supports a war, which is contrary to international law. By acquitting the major from the charge of refusal of obedience and rescinding the demotion to the rank of captain, the federal court indirectly condemns the present federal government, because the unambiguous testimonies given by witnesses during the proceedings indicate that it cannot be ruled out that, by means of the incriminated IT project, the *Bundeswehr* after all *does* support the war the allies are still conducting in Iraq in the year 2003.

The major who had obtained his right had studied pedagogy at one of the two *Bundeswehr* universities. It says much for his character that, at the time, he had come out against the academic staff who denigrated the armed forces by putting up caricatures on their doors that depicted soldiers as pigs giving the military salute. This was symptomatic of the time, when members of the students' movement of 1968 started attacking government authority. While other officer students preferred not to notice this insolence, Z., who then was a second lieutenant, took a stand against it within his faculty. It is evident that in doing so he took the risk of being punished by receiving bad grades intentionally given for academic work that was in fact good. These events, which took place almost 25 years ago, prove that already at that time Z. was a soldier whose sincerity left nothing to be desired by his employer, the armed forces. The soldier acted according to the principles of the *Bundeswehr*'s binding doctrine of *Innere Führung*. He took this military behaviour theory seriously and acted accordingly. It will certainly be correct to say that for him, soldierly values are an integral part of his officer's honour. One must appreciate the lucky circumstances that made it possible that this officer was not psychologically destroyed by his straight professional principles but was confirmed by the supreme administrative court's decision.

Somebody who is able to keep a detached view of the way he is seen by others would have left it at that, and would have been content with this successful court decision, waiting patiently to be promoted to the rank of lieutenant colonel. Z., however, it will have to be said, identifies with his image of a soldier with integrity and does not rest. He seeks every opportunity to proclaim his view that the Iraq War is a war crime and, for example, has to face the fact that high-ranking lecturers on the war in the Middle East know how to cleverly divert his critical remarks about the *Bundeswehr* during discussions. For somebody in the rank of major, it is neither promising nor intelligent to reduce German military policy to one single topic and to urge corrections. Therefore, there is no intention of breaking a lance for this major here.

Rather, the matter to be discussed here is the relationship between obedience and comradeship in a special case. What social and leadership qualities are required to find ways of integrating personalities such as Major Z. into the armed forces such that they both can fully develop their professional abilities and are taken seriously in their criticism of the system? How should the respective superiors of the second lieutenant have dealt with him in order to free him of an inner attitude which specialists refer to as "ethical rigorism"? This form of moral absoluteness may have been adopted by somebody who, for example, has had to use his intellect to assert his rights against an older, physically superior brother from his childhood days on, and actually has found this method to be successful. His parents may have looked on while the younger brother was struggling to find stronger arguments because of his physical weakness. Such family constellations may not be as rare as one might think. What may happen to such soldiers, who quickly become known as difficult, within their unit? There, too, they will have to assert themselves against the imago of the older brother. If they take on the role of the latter, such a background of rivalry may turn into something positive. As military superiors, they can make sure that justice reigns in their area of responsibility. The higher their rank, the larger their sphere of moral influence. Therefore, there are indeed examples of military officers in the rank of general who are highly esteemed by the soldiers under their command because their subordinates feel they are treated fairly. In this context, "fair" does not mean more, nor less, than bringing justified individual expectations into harmony with the objective requirements imposed by military duty. Such a culture of military leadership, which may prevail throughout an entire division and cause the soldiers to dedicate themselves to their profession with heart and soul, may easily be jeopardised or destroyed by political decisions made without consideration for the mature awareness of military values that has developed there. For example, the statement: "Soldiers are murderers!" made by a conscientious objector was deemed an expression of personal opinion by the court concerned, and no punishment was imposed. This was commented by Major General Schultze-Rhonhof as follows: "A state that does not impose punishment for the stigmatisation of its own servicemen as murderers has forfeited the right to call up its youth for compulsory military service." (p. 151). The general, who disagreed with the *Bundeswehr* higher command, described his views later under the title *Wozu noch tapfer sein?* (Why still be brave?). Such events nowadays soon fade into history if they are not made the subject of academic analysis.

In addition to the analysis and evaluation of the reason of a complaint, the aim of administering justice will always be to fulfil the moral requirement of doing justice to the person concerned. In this respect, its objectives overlap with some of those of military pedagogy, which also deals with cases of legal conflict and looks at the underlying problems in order to reflect about possible ways of preventing them from the very start or correcting them once they have occurred. Is a plaintiff now really satisfied with his rehabilitation, or does he continue to feel the urge to try to obtain a further court decision that will do away with his own psychological-social distance between the actual judgement and what he originally had expected it to be?

In the case taken as an example, the plaintiff was given his right by the final court of appeal. His demotion was reversed.

The Decision by the Federal Administrative Court (Bundesverwaltungsgericht, BVerwG)

The competent federal judges explained on 126 pages why the demotion of the major who had refused to co-operate in the development of an IT program had been wrong. They have given him written proof that he has a particular sensitivity in differentiating between good and evil. In doing so, they acknowledged the unbiased and unpretentious way the plaintiff appeared before court. They confirmed his view that a military order is not valid if it violates international law (p. 34). Furthermore, the Federal Republic of Germany was not bound by any kind of obligation to provide military support to the US campaign in Iraq, e.g. as a member of NATO (p. 72). This condition would only have applied if the Iraq War had been based on a reason relating to international law. The massacres Iraq had committed on its own population, e.g. the Kurds, have never been invoked to justify a war, although the moral crime against the civilian population there, including the use of poison gas, had been blatant. There is a quote of a statement by Kofi Annan who had termed the Iraq War by the U.S.A. and its war alliance an illegal intervention (p. 80). Even the German participation in AWACS[1] surveillance missions over Iraq is not justified if it serves fire control functions (p. 95). The federal judges even went so far as to apply the legal principle of "practical concordance" to substantiate that the decision of conscience the major pleaded did not collide with other legal principles worth protecting (p. 108). They finally refuted the fear that their judgement might lead to "massive wear and tear on consciences", triggering a flood of complaints for similar reasons (p. 118). To achieve acquittal, the plaintiff had been able to cite an extraordinary argumentative effort and, above all, convincing reasons.

1 AWACS = Airborne Warning and Control System

What the federal judges criticised was that the superiors of the major had not been able to adequately explain him the situation. Otherwise, they would probably have arrived at the same conclusion as he.

Sharpening the Awareness of the Responsibility in Giving Military Orders

The analysis of the case of complaint the Federal Administrative Court in Leipzig dealt with shows that a precedent was created here. Its inherent challenge consists less in countering possible similar complaints in time but more in fine-tuning theory and practice of *Innere Führung* to the validity and proper ways of dealing with decisions of conscience. Frequently, it is especially the most capable of soldiers, i.e. those who take their professional mission particularly serious, who start having qualms about assuming tasks that border on international law. The case of Major Z. sharpens awareness of the moral sensitiveness of soldiers and situations such as the ones mentioned. It draws our attention to the qualifications of military superiors and legal advisers in the fields of defence policy and military pedagogy. It may be assumed that the case of Major Z. would not have taken on such proportions if he had found people to talk to in his military environment who would have taken his concerns seriously before the actions under disciplinary law took effect. The solution to assign him another task than to develop a software that might be used to support the Iraq War, and if so, certainly in a rather indirect way, could have been implemented already much earlier. What lesson may now be learned from this court decision?

The judgement passed is proof of how, due to the separation of powers in a society whose constitution is based on democracy, the judicial power in this case succeeds in helping a soldier get his due in a delicate, defence police-related conflict situation. The fact that in their statement of reasons the judges refer to the Basic Law of the Federal Republic of Germany and the emphasis the latter places on decisions of conscience makes this opinion of the court an example of how to deal with soldiers and their individuality in a psychologically correct manner. This is becoming important in a global political situation where soldiers must risk their lives as they bear worldwide co-responsibility for peace and freedom. A current conflict has become an example of how the rule of law prevails in the military, which increases the soldiers' confidence in their own actions in possible duty-related conflicts. Additionally, trust in the independence of jurisdiction is fostered, promoting hopes that other national courts will take the same effort in dealing with a comparable complaint and treat it equally thoroughly. If other courts apply similar care in dealing with the culture of the military, applying equally subtle judicial reasoning, they considerably contribute to efforts of military pedagogy at making military ethics turn into a maxim, a guideline of action in international military co-operation. The fundamental relationship of this ethics becomes obvious in the moral correlation between obedience, comradeship and conscience. Anybody who has the right to demand obedience from others and who has to obey orders given by others remains a comrade. It is the soldier's conscience that indicates whether his military superior treats him fairly, while it urges him to act justly towards his superior in deed and in thought. This two-fold moral obligation by superiors and subordinates was illustrated by the example of Major Z.

References

Bundesverwaltungsgericht 2. Wehrdienstsenat 12.04. *Urteil v. 21. 6. 2005 im gerichtlichen Disziplinarverfahren gegen den Soldaten F. P. wg. Befehlsverweigerung* (decision by the Federal Administrative Court of 21 June 2005 upon legal disciplinary proceedings against the soldier F. P. for disobedience to orders). TDG N 1VL 24/03.

Bundeswehr (2006). *Gesetz über die Rechtsstellung der Soldaten (Soldatengesetz).* Berlin: Bundesministerium der Justiz.

Flick, U. (2004): *Triangulation.* Wiesbaden: VS Verlag.

Frankfurter Allgemeine (2006). *Bundeswehr: Ungeheuerliche Indiskretion.* [On-line]. Available: http://www.faz.net/s/Rub594835B672714A1DB1A121534F010EE1/Doc~E5B18885961FB4BB2A57C87DE0327383A~ATpl~Ecommon~Scontent.html.

Hobbes, Th. (1651). Leviathan or the matter, form and authority of government. Stuttgart (1970): Reclam.

Royl, W. (2007). How military training and education are embedded in the societal context. In: H. Annen & W. Royl (Eds): *Military Pedagogy in Progress.* Frankfurt am Main: Peter Lang, pp. 133-151.

Schultze-Rhonhof, G. (1997). *Wozu noch tapfer sein?* (Why still be brave?) Gräfelfing: Resch.

Smith, A. (1776). *Der Wohlstand der Nationen(2008)* (An inquiry into the nature and causes of the wealth of nations). München: Deutscher Taschenbuchverlag.

Gavril Malos, Toma Plesanu & Sorina Mardar

Ethical Code in Romanian Defence Acquisition Domain

Introduction

The core ethical values, especially honesty, integrity, promise-keeping, fairness, pursuit of excellence, and accountability are the building blocks for empowerment and thrust. In Romanian Defence Acquisition Domain is needed ethical leaders who will create organisations in with ethical decision making and ethical behaviors are accepted organisation practices. The gold of ethical leaders and managers is to create organisations in which there is open, honest communication so fear can be replaced with trust. Trust, once established, will enable workers to accept responsibility instead of placing blame and defending against attacks-activities that require huge expenditures of time and energy which impede quality and productivity.

Definition, Purpose and objectives of ethics

During the past few years, the ethical conduct of program managers, defence company executives, defence consultants, contracting officers and others from the acquisition community has been the target of intensive media attention.

We may define ethics in general, as a set of standards of behavior as defined by a group, organisation or society. In practical terms, we define ethics as the standards of conduct which indicate how one should behave with respect to their moral duties and obligations.

Josephon in says etics helps us deal with two aspects: the first involves our ability to distinguish right from wrong, good from evil, and propriety from impropriety; the second involves our commitment to do what is good, right, and proper. Karp and Abramms say the purpose of etics is to translate our value into appropriate behaviours that respond to the realities of day-to-day life.

The implementation of the Etical Code for the personnel with responsibilities in the public acquisition domanin in ROMANIA has the following *objectives:*

1. to establish a set of ethical rules and conditions to protect the public interests during the development of the public acquisition procedures;

2. to perfect the organization structure and the functional relations in order to accurately apply the instructions, norms and procedures regulating the activities within the integrated system of the public acquisition management for Defence and the connected standards;
3. to prevent and eliminate facts that may constitute elements or practices lacking ethics;
4. to make employees conscious of their interest to comply with the Code, thus avoiding the legal consequences that may generate negative effects on them;
5. to protect, maintain and stimulate competition and a normal competitive environment in order to promote the interests of the acquisition beneficiary;
6. to increase the transparency of the public funds usage and of the organization and deployment of the public acquisition procedures.

In our opinion, in order to apply the public acquisition procedures, the employees must comply with the following *principles:*

- integrity, respectively the quality of being integer, honest and incorruptible;
- professional competence and job responsibilities, compliance with professionalism, respectively, competence, impartiality and applying all knowledge, aptitudes and acquired experience;
- free competition and assuring conditions for any product purveyor, respectively, work executor or service provider, irrespective of its nationality to have rights to legally become a contractor.
- public funds usage efficiency and usage of the competitive system and of the economic criteria in order to place the public acquisition contract, respectively;
- transparency and to give information to those interested in the appliance of the procedure for placing a public acquisition contract, respectively;
- equal treatment and to apply without discrimination the selection criteria and those for placing the public acquisition contract, respectively, so that any product purveyor, work executor or service provider should have equal chances to obtain that contract;
- confidentiality and to grant the protection of the commercial secret and intellectual property of the offerer, respectively;
- responsibility and to respect the responsibilities assignment established by law for different factors involved in public acquisitions, respectively, and not to impose some actions according to the detained hierarchical positions.

Factors with negative Influence on Acquisition Ethics

Unfair competition

Contractors will be asked to develop activities in good faith, according to fair practice, respecting the beneficiary interests and the legal competition demands.

It is considered to be contrary to fair commercial practice if the contract is not unilaterally executed or if there is use of some unfair procedure, breach of trust, misdemeanour urge, as well as commercial secret acquisition by tertiary that might know that acquisition implies practices affecting the position on the market of the concurrent merchants.

Given the sanction of that contract nullity and the application of legal sanction measures, the employees who have taken part in any way in the documentation drawing up to elaborate and present the offer or those who are in the assessment committee appointed to place that public acquisition contract have no right to detain the title of associate offerer or subcontractor.

The contractor must be informed about not having the right to employ in any way, in order to carry out the public acquisition contract, natural or legal persons who have taken part in the documentation drawing up to elaborate and present the offer or those who were in the assessment committee appointed to place that public acquisition contract, during a period of at least six months since the contract was concluded. If this condition is not respected, the contracting authority will cancel the public acquisition contract and will take actions to force the payment of indemnification by the contractor and refer the contractor to the ability bodies in order to apply the sanction measurements.

The employees with responsibilities in the public acquisition activity are compelled to:

- respect the confidentiality of the information presented by offerers / candidates in order to comply with the qualifying conditions;
- keep confidentiality on the offer contents as well as on any other additional information requested to the offerer as its disclosing could affect the offerer's right to protect the intellectual property or commercial secrets.

Corruption

Corruption, irrespective of form, active or passive, implies the abusive use of public power in order to obtain an unfair income for oneself or another person, by:

- abuse of power in the exercise of the prerogatives of the public function;
- fraud (cheat and prejudice to other person or entity);
- favouritism;
- setting an arbitrary mechanism of power exercitation in the public acquisition domain;
- inobservance of the provisions for the conflict of interests.

The main proposed measures for preventing and eliminating corruption are:

- setting a system for the employees selection, promotion and assessment as well as a control system, both based on objective recognition of the merits and professional performances;
- Code implementation within all contracting authorities;
- Accurate appointment of the job responsibilities, by means of the job description act, for each employee acting in the public acquisition domain and periodical assessment of their professional activity;

Exclusion will be applied for the offerers who make use of corrupt or fraud practice in the procedure of placing an acquisition contract.

Conflict of interests

The principles that serve as base for preventing the conflict of interests in the exercise of public functions are: impartiality, integrity, decision transparency and public interest supremacy.

After and at the moment of appointment in a function, as well as during its exercise, the employees have the obligation to administrate their interests in such a way that they should prevent any conflict of interests.

The contracting authority personnel must not:

a) be a unique associate or directly take part in the administration or management of any organization or commercial society, excepting the administrative committee of autonomous administrations and commercial societies subordinated to the Defense Ministry, within the Defense Industry or related to it.
b) exercise functions other than those appointed for, excepting the legal plurality of offices, on the terms established by the Defense Minister's order;

If the employee, a member of his/her family or a person who is a relative to him/her up to the level IV inclusively, may have personal benefit after a decision has been taken by the function holder or if that employee has taken part in making that decision, he/she is compelled to report it in written form immediately to superiors and define himself/herself of taking part in such activities.

Those persons who hold public functions in the public acquisition domain will make a declaration of interests according to the law, on their own responsibility regarding their functions and activities in exercise.

Presents and hospitality

It is forbidden for the employees to receive money, loans, presents or other benefits from the part of their official collaborator societies or persons during the acquisition procedure as well as during the contract. Yet, symbolic value gifts may be accepted as representing materials at protocol meetings. Offering this kind of gifts and showing hospitability are perfectly acceptable in commercial relationships. All presents offered, including those with protocol value, must be reported by employees to their superiors. The contracting authority will take measures to have them registered.

Offering hospitability to business partners is a well-known practice for commercial organizations and societies and it is recognized as a fact that this might be a benefit for an efficient behaviour in business world. Lunch or dinner invitations from representatives of some commercial organizations or societies with which the contracting authority collaborates or has intentions of collaboration may be normally accepted as long as the aim is to maintain a fair relationship, and accepting the invitation does not offend the interest of the contracting authority.

The debates with the representatives of these commercial organizations or societies, outside the existing legal background, tend to be considered as deviations from the acquisition ethics.

Those invitations of social nature that refer to sports or cultural events, holidays offers as well as hospitality offers including free of charge boarding and transport must be declined by the contracting authority employees.

Employees must report all proposed hospitability offers, irrespective of them having been accepted or not.

Post-employment

Employees will declare on their own responsibility, in written form, that on their leaving the Defence Ministry they will not develop activities for commercial organizations or societies with which they have collaborated during the exercise of the public function without letting the contracting authority know about it. This liability is held for a period of time of at least six months after they have left the ministry. In order to be hired by such an economic agent in the above-mentioned period of time, the employees are compelled to request the contracting authority's approval. The contracting authority will inform the new employer about the possible incompatibility position that might appear in the relationship with the Defence Ministry.

These rules are meant, when such employment is requested, to grant that:

- there are no reasons to suspect that the actions of any employee of the contracting authority could have been influenced by the offer or promise of a future employment in a certain commercial organization or society;
- no commercial organization or society can obtain an unfair advantage over the Romanian state or its competitors due to the employment of an ex-employee of the contracting authority who had access to commercial secrets by the nature of his/her ex-office.

Conclusions

The bottom line is that within an ethical organization, people will be valued. Valued workers make quality performance a matter of personal pride and honour. In turn, personal pride and honor are the critical ingredients for successful quality organizations.

References

Annen, H. & Royl, W. (Eds.) (2007). *Military Pedagogy in Progress*. Frankfurt am Main: Peter Lang, Internationaler Verlag der Wissenschaften.

Barnes, T.A. (1992). *Logistics Support Training Designed Development*, New York: Mc. Graw – Hill Book.

Blanchard, B.S. (2000). *Logistics Engineering and Management*, New Jersey: Prentice Hall.

Jones, J.V. (1987). *Integrated Logistics Support Handbook*, New York: Mc. Graw – Hill Book.

Josephson, M. (1999). *Making Ethical Decisions*. Los Angeles: Josephson Institute of Ethics.

Karp, H.B. & Abramms, B. (1992). Doing the right thing. *Training & Development, 46* (8), 36-42.

Malos, G. & Sandu I.-E. (2002). *System Engineering for Defence*. Bucharest: Military Technical Academy.
Malos, G. (2005). *System Engineering in Defence Acquisition Management*. Bucharest: National Defence University.

Radomir Saliger & Antonin Müller

Structural Changes at Czech Education System in New Security Environment

Abstract

This article shows a brief historical development of military university education, its current structure and an organizational model of military professional preparation that is realized at the University of Defence in Brno. The presented model reflects on tasks of the development of education conception in the department of defence.

Introduction

The quality of human resources and the level of their preparation (culture and skill) is a presumption of the fighting power of the Czech Armed Forces and their readiness to fulfil the assignments resulting from the interests of the Czech Republic and its membership in the North Atlantic Treaty.

In connection with the foundation of the University of Defence in 2004 emphatic changes in the university preparation of officers were realized. A totally fundamental principle was the decision towards the separation of university education from military training and profession preparation for an achievement of a scope on a relevant and established position.

The implementation substance of this measure projected into a creation of so-called application courses for graduates in bachelor study programmes, their realization was charged to the Military Academy in Vyškov. This was a departmental educational arrangement.

The so-called "Conception of construction" accepted principle determines that the education achieved by a study at the University of Defence must be comparable to civilian university education and accepted in the frames of the European Union. At the same time it must create a background for asserting a soldier after finishing the military service in a frame of a possible second career. To meet this objective, direct measures that concern an organization are required, namely planning, carrying out and the appreciation of the educational process at the University of Defence. Realized educational practice points to the possibilities of the realized model for further evolution of the preparation of future military professionals.

The tasks for the University of Defence result from formulated objectives in the conception of education at the department of defence, namely: a preparation of high-quality 21st century military professionals and the creation of a system that enables the members of this department to achieve qualifying presumptions in the course of their military career.

The main task of military education in the very process of the education of the personnel of the defence department is:

a) an achievement of required special qualification for an exercise of the activities in particular established positions;
b) upkeep of desired special qualification;
c) an enhancement of the culture level in the frame of a lifelong education system;
d) a preparation to a second career.

Czech educational environment

The Czech educational environment has very deep roots. And so has its military offshoot. The first Czech military school was established in 1921, which means tree years after the constitution of the new state Czechoslovakia. (See figure 1). Almost simultaneously, in 1926, a school for medical military specialists was established. The need for education of technicians for the military occurred a bit later, after World War II. – In 1951. This is not yet remote history for us.

The most important event that affected the further development of the Czech educational system was the fall of the Iron Curtain. A partial consequence of the latter was the separation of former Czechoslovakia into the Czech and Slovak Republics respectively as well as the almost simultaneous agreement to reduce military staff in Middle Europe. Three military universities, the educational heritage of a 200'000 men army, were too much for a contemporary modern state with approximately 10 million inhabitants.

This untenable situation had been known for more than a decade, before a total restructuring of the military educational system was prepared and later also realised as a part of a reform of the whole army. Now it can be told in a few words, but the process was extraordinarily tough and sometimes seemed on the brink of not coming to a successful end. But finally it was done and according to the Czech Parliamentary Law N°214/2004, the Czech military university, named "University of Defence", was established. It consists of three faculties and of three institutes. It can be said with a certain degree of hyperbole that it contains the entire good of its three predecessors. Its scheme is shown on figure 2.

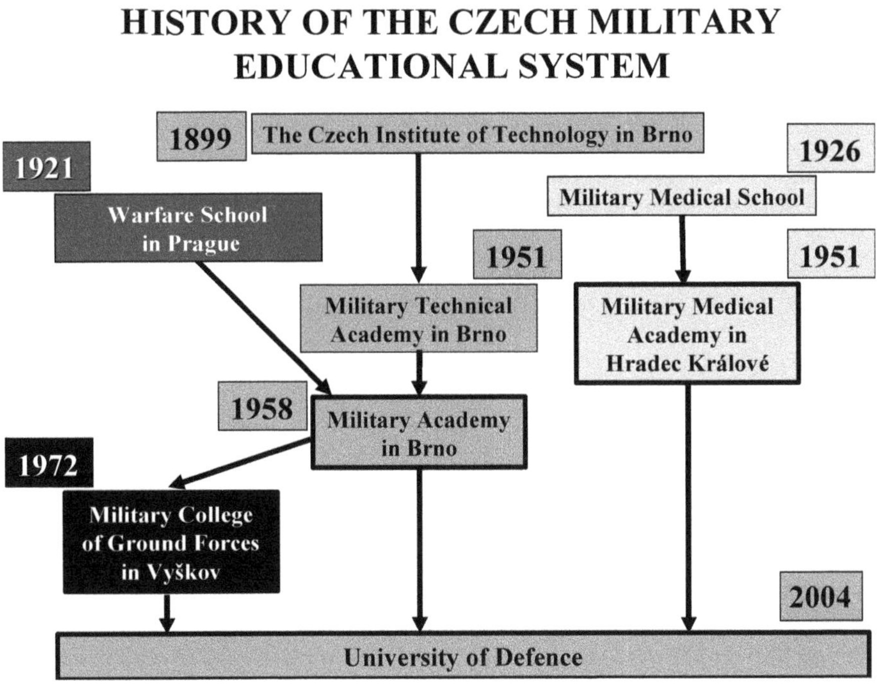

Figure 1. Schema of history of Czech educational system.

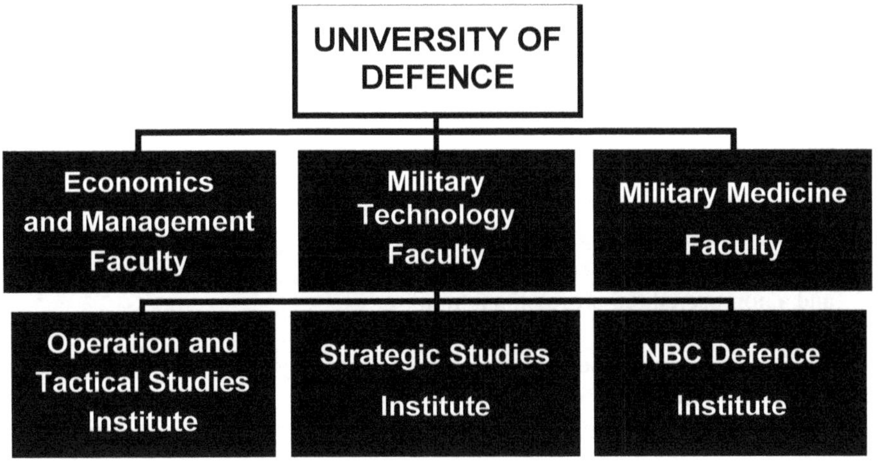

Figure 2. Structure University of Defence

The headquarters of the University is situated at Brno, but the whole school is situated in three cities – the Military Medicine Faculty is located at Hradec Králové and the NBC institute at Vyškov, which means that all three facilities remain in their original cities.

Czech military training and educational system

Of course it is not a Czech invention nor is it a secret that a quality military calls for its own willingness to be built on tree pillars:

- education
- training
- experience.

It is not a coincidence that education was stated at the first position. It is due to its importance in the entire system and – to be honest – partially to the fact that it is our only task in the system. The separation of education from training in the system from the very beginning was one of the most significant changes that happened during the transformation process. The task of the military university is now only to provide the students with university education. I suppose you know that this statement is a bit of a simplification of the real process, but the purpose of this statement is to underline this type of change. The basic military training is not the school's task anymore. For this purpose (and not only for it) a new army facility was established – the Military Academy in Vyškov. Its main tasks are shown on figure 3.

This military facility was established in total accordance with the educational and training resources that had been stated at the birth of the University of Defence:

- Career officers expected to serve for at least 10 years need a minimum studying course duration of 3 years – bachelor degree.
- The new model of officer training consists of an accredited university course and a non-accredited professional training.

Such parameters of the system can be adjusted by the Czech Republic and moreover, its military schools have already made a step towards the realisation of the Bologna Declaration and according to its requirements have divided university education into three consequent, but independent cycles – bachelor, master and doctoral ones. So the University of Defence can organise (and it really organises) all these types of courses.

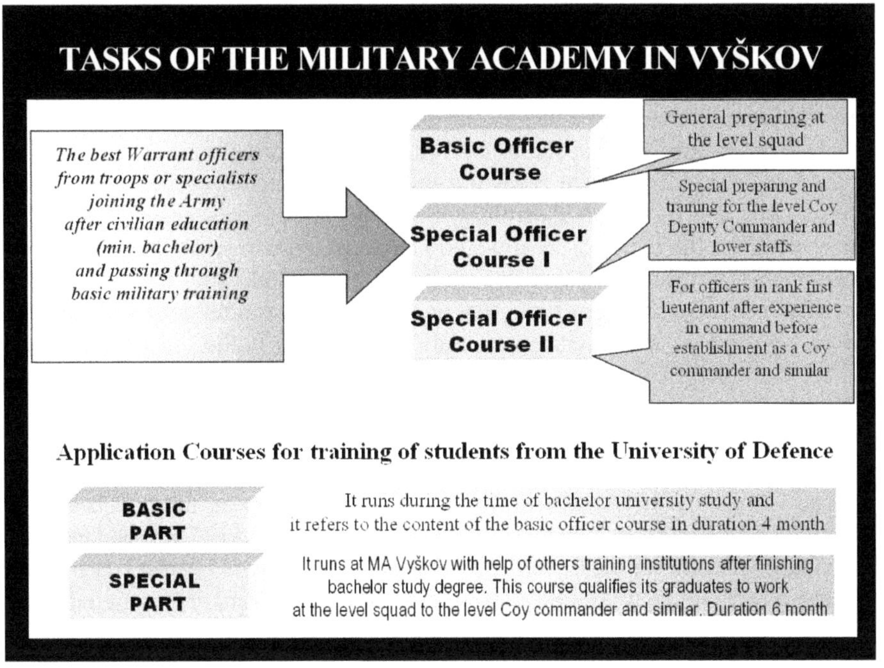

Figure 3. Main tasks of the Military Academy in Vyškov

University level

- Bachelor (Bc.) – duration 3 years
- Master (after Bc.) (Ing. = MSc) – duration 2 years
- Master (M.D., MSc.) – duration 6 or 5 years
- Doctor (Ph.D.) – duration 3 years

Professional military

- Commander courses for senior officers
- Staff courses for senior officers
- Short-term special courses

All these courses offered at the University of Defence, in cooperation with the Military Academy in Vyškov, create a self-contained military educational and training system, in which the officers proceed through their entire military career. Some advance further; some leave the system earlier. It is only up to their ability and to their will to succeed. The scheme of their professional life is on figure 4.

This scheme shows the role of the University of Defence in the Czech military education system and also its contribution to the university educated Czech officer. Fewer than 30 % of the future officers have started their careers outside of the University of Defence. They are specialists in such branches as law, psychology and also for instance theology. Such specialists are necessary in a small amount and for the Czech army it would be uneconomic to teach them ourselves. But a majority of future officers join the army immediately after their A-levels at a civilian school at the age of nineteen and then start a shorter or longer military career. The age of 19 is very appropriate because this is the most suitable age for the formation of life-merits. A military environment surrounds them not only during their professional training but also throughout the ostensibly non-military situation during their university studies. But this indirect effect is usually more effective and stronger than a direct one.

The University of Defence in its new structure shown on figure 2 is able to accomplish all its educational tasks, which it is responsible for. They lead into two directions: inwards for the Czech army to prepare its officers for tasks guaranteeing the Czech sovereignty and outwards for it to safeguard international commitments we are bound with. The most important aspects in order to accomplish these tasks are the development of passive survey systems, NBC identification and defence, primarily against chemical and biological weapons as well as military medicine and pharmacy.

Conclusions

The Czech military education and training system in its new structure of accredited and non-accredited educational courses is able to accomplish all the tasks that are laid in front of it. After restructuring and accepting all the oncoming demands, it now represents a modern structure that is correspondent with the requirements of the third Millennium.

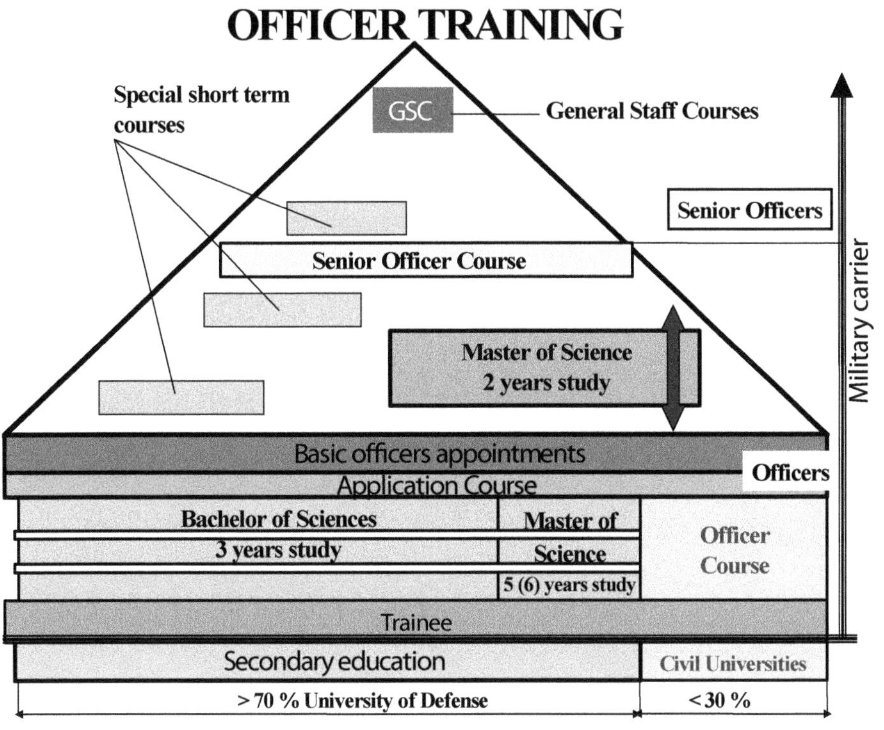

Figure 4. Czech military education and training structure.

References

Čech, V. & Klaban, V. (1995): Personální problémy AČR a nástin způsobu jejich řešení (scientific research VU 200 V-01-21). *Military Academy*, pp. 1-157.

Čech, V., Jevický, J. & Müller, A. (2001). Modelling of the Keeping Process of the University Staff Structure. In: *Proceedings of MOSIS 2001*, Hradec nad Moravicí, May 9. - 11., pp. 9-14.

Koncepce vzdělávání personálu resortu MO, výstavby a rozvoje vojenského školství na roky 2006 – 2011. Praha: MO, 2005.

Lukáš, L. & Velíšek, J. (2005). The *Officers Education System of Czech Armed Forces*, Prezentation at Wroclaw.

3. Applying Military Pedagogy

Victoria Syme-Taylor

Innovative Teaching Methods in Military Pedagogy

Introduction

When looking at new approaches to military education at mid career level it is important to see any course not just as a list of subjects to be covered but to understand why the course has been constructed in a particular manner, how the material links together, and what precisely the student should be trying to achieve in relation to learning objectives and outcomes. This should emphasise the distinction between training, which aims to reproduce a particular task and job, and education which enables development of an individuals intellectual ability. However, this is made on the assumption that military education is an important tool in achieving a professional military. This has not always been accepted when looking at the history of armed forces.

The Challenge

The reason why education is so important to the military officer at this point in his career rest on the nature of the challenges facing defence in the 21st century. While a key strength of military organisations is the solidity and cohesiveness, underpinned by doctrine and systems, it is also true that the challenges posed by the use of military force in the world today require officers who can think and act independently of templated or formulaic guide lines. These challenges flow from the changes in the strategic environment driven by social, economic and political factors. These changes are affecting the nature of warfare and by extension security as a whole. As a consequence, the demand for professionalisation within the Armed Services and particularly at the critical Staff Officer 1 and OF5 rank levels has never been greater.

Defining Aim and End State

We have to be aware of the requirement to have a clearly stated aim and clearly stated end state. In a given course these can be defined as follows:

Aim

To prepare selected officers for high grade appointments, developing their command, analytical and communication skills, and by providing a broad understanding and knowledge of joint, single service and combined operations, and defence as a whole.

Student end state

To have developed a mind that is flexible and can analyse and conceptualise in a military context in order to make timely and logical decisions in all types of subsequent appointments.

Nevertheless necessary questions need to b asked such as; Are these correct? Do they reflect the position of the military in all societies or just some? Do they answer the challenges of the new security environment?

Skill Requirements

The student end state clearly focuses on developing the ability to make timely, logical and military relevant decisions. In more detail, such decisions result from a choice for enactment made from amongst a range of options. In order for the choice to be logical, relevant and effective, it is essential that the issue is properly understood in terms of what is required, and the potential options are interpreted, analysed and evaluated against the required outcome. This is as relevant for the future commander as it is for the future staff officer as staffs that have a genuine understanding of the commanders problem are better equipped to support the commander. All the above requires a student to understand such areas as for example, the international system, the moral dimensions of warfare, the impact of cultural values on societal expectations of war.

The Operational Focus

While the development of problem solving and decision making skills is generic, how those skills are to be deployed is key to the success of any course at this level. A key feature of at and close to OF5 level is the requirement to convert higher level ideas to practical outcomes. The interface between the conceptual and practical is in line with *Joint Doctrine Paper 01* headline description of the operational level. The operational level represents the point at which the issues associated with the practical employment of force are matched to desired strategic outcomes, whether in the field of battle or in the wider management of capability. Achieving that conversion demands effective problem solving and decision making developed upon a firm foundation of knowledge of Defence as a whole. Thus a military course at this level requires an operational focus so that the development of problem solving and decision making skills is professionally relevant. More critically this focus requires more than use of operational level examples in course material. Instead, it requires that students draw out those issues and implications that are relevant to decision making and the achievement of success at the operational level.

For example students will need to see how strategic thinkers develop their ideas on war and how they can or cannot be applied to the modern context. This is also the case when looking at historic examples of leadership qualities. Case studies illuminate for the students those key attributes and indeed key failings.

Learning

Drawing on the analysis above, it can be shown that the primary learning aims is to develop professionally relevant and effective problem solving and decision making skills. Professional relevance is achieved through focusing course material at the operational level and this requires students to consider the following factors as constant and central to learning.

1. *Command and Leadership*. Here the student needs to understand the obstacles that may exist in a given situation. These obstacles may be cultural, organisational, based on perceptions of trust and confidence or a lack of understanding, due to physical and resource constraints, or most likely a combination of all or some of these. The aim therefore is to encourage systematic and reflective understanding of the issues and challenges confronting commanders and leaders working at the operational level, in order to promote personal and team Command and Leadership qualities.
2. *Key Questions* Success at the *operational* level demands the ability to identify key issues amongst a mass of detail. While and increase in knowledge may enable identification of a wider set of questions, effort has to be made to tackle the decisive issues. This requires great clarity of thought and the aim is to develop the learner's ability to define the critical decisions required for success at the operational level.

3. *Practical Implications.* The aim *under* this heading is to enable analysis of operational level problems and the development of appropriate organisational, methodological and resource based solutions.
4. *Solutions.* The challenges and problems confronted at the operational level are invariably situation specific and often unique. While taught processes and procedures may help to organise the development of solutions, unique problems require independent, innovative and creative thinking from the outset if the solutions are to be effective. Where appropriate that thinking may also need to be genuinely original. The aim under this heading is to develop the learner's ability to evaluate the implications and issues for the operational level that arise from the strategic context and tactical realities.

Teaching Level and Methods

The above requirements and methodology are focused at the post graduate level that is the requirement to develop intellectual ability on top of enhanced knowledge base. The following sets of complementary teaching methods are used.

1. Imparting and enhancing knowledge through methods such as small group activates, media (e.g. video), lectures, research (essential and wider reading) and e learning.
2. Active learning including practising problem solving and decision making through methods such as syndicate room discussions, debate, brainstorming, exercises, case studies, role-play and student led presentations.

Assessment

The assessment of a course at this level (postgraduate) reflects best practice in UK higher education. Students should be aware of clearly defined objectives and outcomes. Assessment should also be reflective in its approach. Examples of assessment include essay, debate, communication and team leadership skills. Students will also have academic assessment reflected in their end of course report.

Innovative Teaching Methods and Approaches

Students on Military courses must expect to be challenged by the course material available to them. As such we as educators must seek to counter the criticism that we allow a stereotypical approach to military education to develop. Thus sources of research material made available tot eh students must come from a diverse range of sources and reflect a wide spectrum of opinions. Students should at his level be able to seek out appropriate evidence to back analysis and be they innovative in the research material they use. The following example of a course taught as part of the MA in Defence Studies will show how this can be achieved.

Images of War

This course looks at how conflict is expressed in popular and elitist mediums. Memoirs, diaries and poetry are used to explain individual responses to conflict. Students are asked whether these accounts affected the way future generations viewed warfare. The requirement for memorialisation after war is seen in the context of a need for a collective explanation for sacrifice. The role of war in re enforcing national identity is examined using propaganda images, art and films. An analysis of war films post conflicts reflect the use of war as a medium for human emotions and in some cases to re enforce the "never again" perception of war. The subject is, by its nature, multidisciplinary in approach and uses diverse mediums. It requires from the students a degree of textual analysis. Witnessing military officers reading out poetry is rewarding! At the end of this period of study students should be able to assess how war continues to hold a central role in the national psyche and how the experience of past conflicts affects the way nation's states view conflict. The course is taught in syndicate groups of 12. Students are required to present on a chosen topic and write an essay of 6000 words.

Conclusions

In conclusion, it can be observed, that military education of officers at mid career revel must be continuously reviewed for two reasons. Firstly, a requirement to reflect the changing nature of the security environment and secondly a need to reflect changes in best practice in higher education. The particular intellectual needs of military officials must be seen as unique, but it is of vital importance that the education they receive is of the highest standard and comparable to the post graduate level offered to civilian students.

Teaching has to be seen to as innovative with an emphasis placed on differing learning processes. Students must also be exposed to information and material that goes beyond the traditional assumptions of what constitutes "military education".

All this enhances the student's ability to make accurate and balanced decisions as commanders or as a support to command decisions.

David Hall

Clio and Mars on Tour. Applied History – Battlefield Tours and Staff Rides in contemporary professional military education

'... Crafty men condemn studies; simple men admire them; and wise men use them ...'
(Francis Bacon, 1597)

Introduction

Battlefield tours and staff rides have been an established concept in the formal education and training of British officers since the late nineteenth century.[1] When Major-General (later Field Marshal Sir) Henry Wilson was the Commandant of the Army Staff College before the Great War, he was an incongruous feature on the battlefields of the Franco-Prussian War – fit, flamboyant and controversial, his lanky frame perched atop a bicycle, with an assortment of classified materials bulging from his saddlebags. In his day, the administrative instruction for the staff ride was simple: officer/students were told to go to Thomas Cook's, purchase a train ticket to Metz and put up at a decent hotel when they got there. Dress, of course, was 'serviceable country clothes and sturdy footgear'. A quick count of the high proportion of either Barbour or shooting jackets as well as Hunter wellies worn by British officers on battlefield tours and staff rides today does suggest that some things have not changed.

The terms battlefield tour and staff ride are often seen as being one and the same and, as such, are often used interchangeably. Whilst there are some features that are common to both they are in fact very different activities. A battlefield tour is really a trip around a former combat zone with commentary provided by a veteran, a historian, or an experienced guide. It also requires very little if any preparation and study by the participants. A staff ride, however, is something more. It is an educational method of teaching command and leadership, planning, decision-making and the conduct of operations, and in order to succeed it requires

1 Brian Bond, *The Victorian Army and the Staff College 1854-1914* (London: Eyre Methuen, 1972) pp.175-177. See also David Ian Hall (ed.) *The Relevance and Role of the Battlefield Tour and the Staff Ride for Armed Forces in the 21st Century*, SCSI The Occasional Number 48 (Strategic and Combat Studies Institute, Shrivenham, UK, 2005)

significant preparation by and active participation from all of the participants on the ride.

On a battlefield tour the historian/guide will answer the questions 'what happened?', 'why?', and 'with what result?', but on a staff ride the process goes much deeper and further. Either working individually or in syndicates, the participants first establish a factual narrative of the individuals, actions and events under examination. They then consider whether or not the commander made the right decision (based on the information available to him) at the appropriate juncture. Alternative courses of action are discussed and critically analysed and then the officers have an opportunity to draw on their own contemporary experiences, doctrine and technology for example, to question how they would have acted in his place if presented with a relatively similar situation. In this respect, a staff ride can, therefore, resemble a management school case study conducted *in situ*.[2]

Even in General Wilson's day, it was clear that a staff ride was not simply a battlefield tour but rather an effort to use the past to help illuminate the present. Thus, when Wilson's officers ran the route of von Bredow's cavalry charge on the battlefield of Rezonville (16 August 1870) they were not necessarily convinced that the cavalry of their own day could repeat similar victorious feats but rather they were struck by the way in which this bold stroke helped paralyse an already hesitant French commander. The lesson they identified was how a simple tactical action had effects which were operational and strategic – whose consequences, in their way, have endured till our own lifetimes.[3]

Advancing forward approximately 100 years to May 2004, just before the official ceremonies to mark the 60th anniversary of D-Day began, I had the privilege of being a consultant historian on a NATO HQ's staff ride in Normandy. The HQ of the NATO Rapid Deployable Corps – Italy (NRDC-ITA) examined the D-Day invasion and a number of the key battles in the Normandy campaign. Their stated aim was '*to provide through military historical example, a practical focus for the study and analysis of high command and the planning and conduct of future campaigns and major operations, within a combined and joint context.*'[4]

---

2 Cranfield Security Studies Institute, *Battlefield Tours and Staff Rides* [nd] pp.13-14.

3 Bond (note 1) pp.248-250; Richard Holmes, 'Foreword' in the Higher Command and Staff Course (hereafter HCSC), *Exercise STAFF RIDE Reference and Study Guide* (Joint Services Command and Staff College (hereafter JSCSC), Shrivenham, 2006) p.iii

4 NATO Rapid Deployable Corps Headquarters (hereafter NRDC-ITA), Exercise EAGLE TOUR 2004 (11-16 May 2004) Pocket Reference Guide and Supplementary Reader (Solbiate, Olona, Italy, 2004)

EAGLE TOUR 04

EAGLE TOUR 04 was the exercise name given to the NRDC-ITA battlefield study tour and it was aimed at both senior and junior staff officers. In total, eighty officers from this multinational HQ took part, including one Lieutenant-General and two Major-Generals. For three very active days, we travelled around Normandy enthusiastically engaged in 'applied military history'.

The itinerary included a number of the familiar and a few of the not so familiar sites:

- Pegasus Bridge (site of the glider landings by Major John Howard and the 2nd Ox & Bucks Light Infantry in the early hours of 6th June 1944 – an operation which Air Marshal Sir Trafford Leigh-Mallory described as 'the finest flying in the war').
- Sword Beach (Queen sector, where the 3rd British Infantry Division landed on the morning of 6th June).
- German Strong Point "Hillman" (the regimental headquarters of the 736th Grenadier Regiment under Colonel von Krug, located behind Sword Beach).
- Arromanches (the site of *Port Winston*, the British Mulberry harbour).
- Longues Battery (a major German Coastal Battery, part of the Atlantic Wall defences).
- Omaha Beach (where the 1st and 29th US Infantry Divisions landed and suffered the highest number of casualties of all the allied assault forces on D-Day).
- Pointe du Hoc (which was assaulted and occupied by the US Rangers again with very high casualties).
- Breville (overlooking the Line of Departure for Operation Goodwood, (18-20 July 1944), the British and Canadian armoured attack east of Caen beginning on 18 July).
- La Petite Chapelle (Mortain – where allied air power, mainly RAF and RCAF Typhoon rocket firing fighters defeated the German counter-attack, Operation Lüttich, beginning on 7 August).
- Ste Lambert sur Dives (where Major David Currie of the 29th Canadian Armoured Reconnaissance Regiment won a Victoria Cross in bitter fighting in the Falaise Pocket on 18 August 1944).
- and lastly, the Polish Memorial at Mont Ormel (where the Polish Armoured Division fought to seal the neck of the Falaise Pocket in August 1944, thus completing the encirclement of the German forces in Normandy and ending the campaign.[5]

5 NRDC-ITA, Exercise EAGLE TOUR 2004, Pocket Reference Guide, p.3

I have described the outline of the NRDC-ITA staff ride in some detail because it was based on the Normandy component of the very successful British Higher Command and Staff Course (HCSC) staff ride. The HCSC staff ride is run annually on a fully joint basis and over the last few years it has established itself as the premier staff ride conducted in the British armed forces today.

One slight departure from the HCSC Staff Ride format was the desire of the HQ NRDC-ITA to hold formal remembrance services, including the laying of a wreath, at a number of the various national cemeteries and memorials in Normandy. The NRDC-ITA visited the Commonwealth War Graves Commission Cemetery in Ranville (near Pegasus Bridge), which contains some 2,151 graves. They also visited the German cemetery at La Cambe (21,300 graves) and the American Cemetery behind Omaha Beach at Colleville-sur-Mer (9,387 graves). A wreath was also laid at the Polish Memorial at Mont Ormel. At each national site an officer representing that nation gave a short speech before a wreath was laid, a prayer said and a minute of silence observed.

Not surprisingly, perhaps, all of the speeches were very different yet equally appropriate for the place and the occasion. One speech, however, struck me as being more poignant than the others. A young attack-helicopter pilot, with considerable recent combat experience, delivered a speech that was truly remarkable for both the humanity and the humility that it conveyed. She began by drawing the group's attention to the row on row of pristine crosses that seemed to stretch beyond the horizon, reminding us all that each one marked a young life that had ended prematurely because of war. Each cross, she continued, also represented a mother's loss, and the grief and loss that grew outwards, seemingly exponentially, through family, friends, community and nation. The cost of battle is never cheap, which is why, she concluded, a Headquarters' Staff, like the one in which she was a member, tasked with planning and conducting future operations, had an enormous responsibility to 'get it right'.

For this young American officer and her NATO colleagues, the Normandy Staff Ride met their desired aims and objectives. In a brief yet very practical way it helped them to gain a broader understanding of the nature of high command. It enhanced their ability to identify and focus on key strategic, operational, joint and combined issues, and it revitalised their professional interest in military history. These are the three higher-level training objectives of the HCSC staff ride.

The NRDC-ITA staff ride had an additional unplanned for benefit that nonetheless received considerable encouragement on the ride itself – a comprehensive discussion on interoperability.[6] At first the questions focused on the conventional

6 The standard definition of interoperability – the one used by the military establishments of the NATO allies – is the 'ability of systems, units or forces to provide services to

and expected topics of weapons and communications systems, new technologies, and the ability or otherwise of different units and Services to operate effectively together. These time-honoured concerns, however, quickly gave way to a more sophisticated line of questioning that focused on cultural and societal obstacles to effective co-operation between the different Services and alliance partners. Left unattended, the sources of behavioural differences can create havoc and be much more destructive to effective co-operation than any number of differences in the technical realm of systems (weapons or communication), units or force structures.

Objective

The aim of this short article is to generate a discussion on the value of history and more specifically applied history – in the form of battlefield tours and staff rides – to modern military education and the professional development of serving military officers. I would also like you, my fellow students and scholars of military pedagogy, to consider whether it is possible to evaluate the effectiveness of this type of education/training and, if it is, how to do so? Accepted and standardised forms of measurement such as benchmarking and the acquisition and analysis of empirical evidence tell us little and are mostly ineffective. On the other hand, most officers I have had the pleasure to take on Staff Rides think that they are very useful. But how do we know if they are or they are not? And if they are, how do we keep them relevant in a fast paced and ever changing security environment?

Are battlefield tours and staff rides just good fun – unit bonding and, perhaps, 'bottlefield' holidays at the Sovereign's or government's expense – or do they add to a commander's professional development, better unit cohesion, improved inter-Service understanding and even enhanced appreciation and understanding amongst allies and coalition partners (as was the case in the NRDC-ITA staff ride)? What questions do we need to ask, and answer, to be sure that the time and resources devoted to these types of educational/training activities are being used wisely and that the desired objectives are achieved?

---

and accept services from other systems, units or forces and to use the services so exchanged to enable them to operate effectively together'. See The North Atlantic Treaty Organisation, NATO Standardization Agency, *NATO Glossary of Terms and Definitions*, AAP-6(2002), 2-I-7

History of British Staff Rides and Battlefield Tours

The modern concept of the historical staff ride has its roots in Prussian/German traditions beginning with Frederick the Great, who walked the ground on which past battles had been fought and considered tactical alternatives to the decisions taken by past commanders. At the end of the Napoleonic Wars the concept of combining theoretical instruction and practical exercises in the field was refined first by Scharnhorst and Clausewitz and again later in the nineteenth century by Helmuth von Moltke, the elder, Chief of the Great General Staff from 1857 to 1887.

Under von Moltke's direction, the long-standing Prussian tradition of officers reading military history and then taking their theoretical studies outdoors in the form of staff rides and war games was institutionalised. The rationale for studying military history seemed obvious: it provided 'lessons for the future'. And staff rides were an effective method of testing both ideas and individual officers destined for high command appointments. With Prussia's victory over France in 1870, the staff ride concept was copied by other European nations.

The exact date of the first officially sanctioned British staff ride or battlefield tour – as a part of an officer's professional education and training – is not entirely clear. But staff tours – both overseas and to a much greater degree in the UK – were a semi-regular feature in training late Victorian Army officers during the 1890s. Not surprisingly, overseas tours studied the battlefields of Prussia's victories in 1870.

Following the First World War, British officers on the Camberley staff course went to Flanders and studied Haig's battlefields.

The vast majority of battles and campaigns studied by the British services are within a few hours drive of Calais, the most fought-over corner of Europe. This is illustrated rather prominently by one Royal artillery officer, who when looking for billets in North-East France during December 1939 found:

On the edge of the village stood a small farmhouse in a very ruinous condition. It was utterly deserted, and obviously hadn't been tenanted for years. I opened the door, which led straight into a room. Chalked in big letters on the opposite wall was the command:

WIPE YOUR FEET, PLEASE 1918

It was so peremptory that I almost obeyed, forgetting that it was not 1918 but 1939, and that only ghosts of a past officers' mess in a past war would grouse if I didn't. It stood there exactly as it had been left after the Armistice. I could almost imagine that, with terrible clairvoyance, its occupants had foreseen that it might be wanted again. In a corner were a rusty bayonet and a Mills bomb. The stained deal table and three wooden chairs were thick with the dust of twenty-one years

of peace .... There was something pathetic about this little heap of forlorn-looking debris, a legacy from one British army to another...
(in: J.C. Austin, *Return via Dunkirk*, 1940)

Shortly after the Second World War ended, Camberley students as well as the British Army of the Rhine (BAOR) began visiting areas of North West Europe, Italy and North Africa where significant battles of the Second World War had been fought. British staff rides also resumed under Field Marshal Montgomery's guidance – studying his victories – and often included veterans and former commanders, such as Lt General Sir Colin Barber, Colonel Pine-Coffin, Major Howard, Canadian General Guy Simons, and German Colonel Hans von Luck, formerly of 21st Panzer Division. The last such ride was held in 1979, ending because of the fragility of the guest speakers. To call them staff rides, however, is misleading. In essence they were tours, conducted at a gallop and with little if any advance preparation.[7]

Camberley resurrected the 'staff ride' in 1988 as part of the Army's Higher Command and Staff Course. HCSC officers researched a particular aspect of one of the campaigns or battles examined on their staff ride, produced a preliminary research paper, and led on a question that was tied to a particular stand. Applied history was back on the Army's syllabus. But this was not just academic history under examination. Students on the HCSC were also tasked with identifying and commenting on a contemporary 'pull-through' (an aspect of contemporary relevance) that embraced the military strategic and operational levels of war. To a degree, as with the Prussian staff ride concept, they were examining the past for lessons that had applicability in the present and possibly the future.

With the foundation of the Joint Services Command and Staff College (JSCSC) in 1997, reflecting the inevitability of both Combined (multi-national) and Joint (tri-service) operations for the British military, the HCSC Staff Ride was broadened to include greater emphasis on the planning and conduct of air and maritime operations. As in the past, Operation Overlord and the Normandy Campaign provided the richest training ground for this combination of theoretical and practical study.[8]

7 Archive of the JSCSC (hereafter JSCSC), *Staff College Battlefield Tour 1947-1979* [nd]; and JSCSC, Kenneth White, 'The Battlefield Tour: An Historical Memoir' [nd] pp.1-5

8 JSCSC, HCSC, *Exercise STAFF RIDE*, Reference and Study Guide for the years 1998 to 2006

Why Normandy (D-Day and the Normandy Campaign 1944)?

First, Normandy is the setting of the last great set-piece battle by the Western Allies in the Second World War – a large-scale airborne and amphibious landing operation which offers numerous lessons for all of the functional branches of the armed forces today, nationally and in coalition. Second, the Normandy battlefields today are largely similar to what they were sixty years ago and they are easily accessible both in terms of physical access and financial costs. Turning to pedagogical reasons, the D-Day landings and the successive battles that make up the Normandy Campaign also provide contemporary military commanders and their planning staffs with numerous case studies on Combined (multi-national) and Joint (air, land and sea) operations and campaigns. Operation Overlord was the product of all arms landing forces, naval forces that ranged from battleships to landing craft and submarines, strategic and tactical air forces as well as the largest assortment of transport aircraft ever to participate in a single battle, the Merchant navy, significant contributions from civilians and industry, and decisive input from the diplomatic and the Intelligence services. Last and not to be overlooked when understanding why the British, American and Canadian armed forces study the Normandy campaign, they won!

HCSC Staff Ride

The HCSC staff ride embraces applied history, drawing on historical illustrations to highlight contemporary issues, to provide a practical focus for the study of high command and the planning and conduct of future campaigns and major operations within a combined and joint context. The ride itself is a ten-day Marathon across northern Belgium and France with a significant part of it taking place in Normandy. It is run annually for a maximum of twenty-four senior officers drawn from the three Services. Most rides also include three senior officers drawn from key NATO partners as well as two or three civil servants of relatively equal grade to their military colleagues. The specific training objectives of the HCSC staff ride may be summarised as thus:

- To examine the military-strategic and operational-level dimensions of command, decision making and resource allocation.
- To explore the application of the principles of war and practise of operational art during the campaigns, major operations and battles under study.
- To assess the impact of personality, through the study of senior personalities, on the conduct of war.

- To analyse the significance of environmental and component factors in the planning and conduct of major operations and battles.
- To stimulate the conduct of original research, expanding student/ officers' knowledge of military history at the strategic, operational and higher-tactical levels of conflict.
- To assess the relevance of lessons derived from military history for operations today.
- To examine the impact of technology on the conduct of war.[9]

In terms of its approach, the HCSC staff ride involves three stages of preliminary study, field activity and integration in some form of after action review. Each officer/student is required to write a 5,000-word research paper and then present their essay topic at the corresponding stand on the ride. A former Deputy Director of the HCSC has noted that 'the combined pressures of personal pride and peer review normally generate some excellent papers as well as searching questions and stimulating debate on the stands.'[10] In addition to meeting specific training objectives, the HCSC staff ride also aims to achieve long-term educational benefits by encouraging personal professional development through awakening or reinforcing an interest in military history and offering an analytical framework for the systematic study of campaigns and battles. The Directing Staff and the three Course historians (air, land and maritime specialists) are an integral part of this educational process throughout.

Once on the stands, a member of the Directing Staff provides a short orientation, noting directional and topographical features as well as any key military deployments, force structures and command issues. A five-minute tactical/historical narrative by one of the HCSC historians follows. Three academic historians support the HCSC, each one specialising in one of the component environments – air, land or naval warfare. The student/officer then has approximately twenty minutes to present their essay topic and identify any themes relevant to contemporary military issues. The academic historians then field any questions of fact before a general discussion concludes the formal activities at the stand.

At the end of each day's exertions, a 'hot-wash' discussion is held at the overnight hostelry. The aim here is to capture the main points that emerged from the day's presentations and discussions, and to articulate any enduring themes of contemporary relevance. A slightly longer and more formal review is held at the end of the staff ride.

9 JSCSC, HCSC, *Exercise STAFF RIDE*, Reference and Study Guide 2006, p.1

10 From a series of discussions the author had with Brigadier R.A.M.S. Melvin, JSCSC, Autumn 2001

Beyond the Normandy beaches

Using similar educational and training objectives as the HCSC, a growing number of British headquarters, military colleges and training establishments, and other air, naval and military units, have embarked on battlefield study tours and staff rides to analyse historical battles and campaigns in order 'to highlight lessons applicable in the Current Operating Environment (COE)'.[11] They have also gone further afield, well beyond the beaches in Normandy, in their search for relevant campaigns that highlight joint and combined operations as well as urban and asymmetric warfare. In the past three or four years large and small groups of British sailors, soldiers and airmen have visited *Sicily* (to study the Allied joint and combined campaign Operations Husky), mainland Italy (to study Operation Shingle, the Allied amphibious landings at Anzio in January 1944; and the 1944 battles to break through the Gustav Line at Cassino), Russia (Moscow to study Napoleon's 1812 invasion of Russia and the battle of Borodino; and Volgograd to study *Case Blau* the battle of Stalingrad); the Falkland Islands (to study the 1982 conflict between Britain and Argentina); Sedan (to study Blitzkrieg and the German invasion of France in 1940); and Berlin (to study the Allied Strategic Bombing Campaign; the Soviet Red Army's victory in 1945; the Berlin Airlift; and the Cold War). The HQ of the NATO Rapid Deployable Corps – Italy (NRDC-ITA) has also built on the success of Eagle Tour 04 and run further historical battlefield study tours to Sicily (Operation Husky) and north-east France and Belgium (Blitzkrieg – the German Invasion of France and the Low Countries in 1940). Eagle Tour 08, conducted in the spring of 2008, focused on the German airborne assault on Crete in 1941. The battle for Crete offers the contemporary military officer many areas of study in conventional joint and combined warfare as well as less conventional civilian resistance to an invading army and the impact of harsh reprisals administered by the occupying forces. It is also one of the great asymmetric battles of the Second World War.

The current surge of interest in battlefield study tours and staff rides has also led to a number of previously unexplored areas of study. This is most obvious in the 'cognitive domain', which for a growing number of academics and military theorists in the defence community is the dominant domain in war. Battlefield study tours and staff rides offer stark illustrations of the inherent difficulties in obtaining situational and shared situational awareness and understanding – themes that have generated a great deal of discussion within the defence community. 'Common Operational Pictures', 'Common Relevant Operational Pictures', 'Common

11 Headquarters Land Warfare Centre, *Exercise Case Blau*, Reference and Study Guide, 2007. See also 17 Regt 51 Sqn Port and Maritime Regiment, Royal Logistics Corps, *Exercise Seahorse Shingle*, Admin Instruction and Study Guide, June 2007; and Royal Naval Division (RND), JSCSC, *Exercise Staff Ride Sicily*, Study Guide, 2007

Relevant Actionable Pictures' and the grandest of them all, 'the Operational Net Assessment' cannot be achieved merely through real-time access to a common set of comprehensive data. Achieving cognitive commonality is a philosophical problem not a technical one. After all, war is a human activity and the myriad of problems associated with war-fighting require human-centric rather than technical solutions. Time and time again, this basic lesson can be seen played out on the battlefields of the past.

Conclusion

Battlefields offer a snapshot of the past for they can illustrate almost any principle of war or lesson at the strategic, operational or tactical level of war, and transport, as if in a time machine, the officer/student back in history. It can certainly be argued that the issues that battlefield study tours and staff rides address transcend time and place and offer a short-cut to equipping Service personnel with an understanding of the 'realities of war'. The horror of combat and the challenges of leadership in a fast-moving, uncertain environment can be learned in a classroom, but so much value is added by studying these issues on the very field of battle where they were played out. Every mistake or hesitation may have cost lives; the evidence usually lies not far away in a war cemetery. Nor, in today's joint and combined environment, does the concept remain land-locked. The USAF has been experimenting with an 'air ride': study the planning for a wartime strategic bombing mission, fly the mission (without the bombs), land and inspect the target area. Staff rides need not, however, be restricted to an examination of combat operations, command decisions, and tactics. Reflecting on the new comprehensive approach doctrine, combining military with non-military agencies, particularly in the post war-fighting stage of contemporary operations, there is significant scope for staff rides that explore non-kinetic activities, such as Phase Four operations, as well as cultural and societal interoperability challenges.

The staff ride is a proven educational construct that has evolved over the last two centuries. It is also a fairly flexible training vehicle, and it can be designed to achieve many training objectives. When used well, battlefield tours and staff rides offer unparalleled opportunities to prepare Service personnel not only for war but also for the many diverse challenges of contemporary operations. Whether it is a conventional conflict or an asymmetrical opponent there are relevant historical models worthy of reflection and study. And with a wider-than-ever range of threats and challenges the military have to encounter today, military history and through it, battlefield tours and staff rides offer past examples of every conceivable situation, from blitzkrieg and military sieges (Basrah and Baghdad have reminded us how the medieval concept of the siege can return to haunt us) to psychological or media operations and civil affairs. It certainly seems that this 19th Century teaching device still offers much that is relevant to military education and professional development in the 21st Century.

References

Bond, Brian (1972). *The Victorian Army and the Staff College 1854-1914.* London: Eyre Methuen.

Caddick-Adams, Peter (2005). Footprints in the Mud: The British Army's Approach to the Battlefield Tour Experience, *Defence Studies, 5*(1), 24-25.

Hall, David Ian (2005). The Modern Model of the Battlefield Tour and Staff Ride: Post-1815 Prussian and German Traditions, *Defence Studies*, *5* (1), 37-47.

Maltz, Richard (2007). Shared Situational Understanding: A Summary of Fundamental Principles and Iconoclastic Observations', *SITREP: A Publication of the Royal Canadian Military Institute, 67* (5), 6-9.

Melvin, R.A.M.S. (2005). Contemporary Battlefield Tours and Staff Rides: A Military Practitioner's View, *Defence Studies*, *5* (1), 71-72.

Can Nakkas & Hubert Annen

Portfolio Assessment – Achieving Sustainablility in Officers' Education

Of the three ways in which men think that they acquire knowledge of things – authority, reasoning, and experience – only the last is effective […]
(Roger Bacon 1214-1294)

Introduction

The classical approach of ensuring that students retain and memorize the subject matter is to let them write a test at the end of semester. Such tests can differ in structuring and normally consist of a battery of questions, ranging from multiple choice to open or closed response questions (an essay may be viewed as a distinct case of an open response question). The advantage of such an approach on the one hand is the possibility to determine – and to a certain degree influence – whether a student has learned the subject matter, on the other hand it allows the teacher to evaluate and mark the student's performance and consequently to compare it with that of the other students.

In study courses and academic fields that mainly rely on processing information this kind of assessment may be adequate. When we look at educational areas, however, that teach values and focus on providing students with soft skills, this approach has its shortcomings. As the objective of these educational fields is to cultivate in students a characteristic trait or mindset and to inseminate them with an attitude or a mentality, respectively, the technique of cramming one's brain with disparate nuggets of knowledge alone will not suffice. Or to state the problem bluntly (Annen, 2007):"…students only take in what appears relevant for the examinations; contents are learned mechanically, processed specifically for the examination and then quickly forgotten again."

Normative knowledge as objective

This problem leads us to two important topics in pedagogy and didactics, namely epistemology and sustainability. While the former relates to the difference between *descriptive* (i.e. positive) knowledge, *procedural* (i.e. instrumental) knowledge and *normative* (i.e. prescriptive) knowledge, the latter relates – in this context – to the question of how knowledge actually can be transferred into practice and thus becomes effective. Both topics are inasmuch related, as they have an influence on instruction and testing.

Descriptive knowledge describes the nature, properties or state of an activity, thing, process or system and includes statements of causation, comparison, explanation, observation, prediction and relationship. Procedural knowledge (i.e. "know-how") describes how to do something, including methods of inquiry, and criteria for using skills, algorithms, techniques, and methods. Normative knowledge, however, expresses a judgment on the importance, morality or value of an activity, thing, process or system, and it includes judgments of good and bad, right and wrong, better and worse; as well as statements about how an activity, thing, process or system should or ought to be (Feldman, 2003).

When we look at military pedagogy, it becomes perfectly clear that although descriptive knowledge constitutes an essential part of an officer's curriculum, the main aspect in teaching e.g. military ethics, cultural awareness etc. is that it demands first and foremost an expertise in normative (and to a lesser extent) procedural knowledge, while nevertheless having a stabilizing effect and ensuring the functional order of the system (Annen, 2002).

Instruction drives testing

Since the emphasis of military pedagogic training lies mainly on imparting prescriptive (and instrumental) knowledge, and the objective of military pedagogic courses e.g. in ethics obviously are intended to be practice-oriented *practice-effective*, respectively, pencil and paper tests will only insufficiently assess – and much less contribute to – sustainability of such education. Instead of merely assessing outcomes, the focus should be laid on processes. As a consequence, in order to assess and also contribute to sustainability one might need to reconsider the old pedagogic precept of "testing drives instruction" and conversely let "instruction drive testing" (Annen, 2007). This would mean that the internal processes themselves, which are required to successfully learn and internalize military pedagogic teachings, would have to be represented and assessed by the evaluation method. These internal processes include e.g. responsibility, self-reflection, self-initiative and self-criticism. One assessment tool that actually incorporates all of these characteristics is the portfolio.

Portfolio

What constitutes a portfolio? In short, a portfolio is a *purposeful collection of student work exhibiting his efforts, progress, and achievements in areas of the curriculum*. The collection must be based on the student's participation in selecting contents, and include criteria for selection; criteria for judging its merits; and most importantly evidence of the student's self-reflection. It should represent a profile of a student's work and efforts; student-selected samples of work experiences related to those objectives being taught and assessed; and documents exhibiting the student's growth and development toward mastering the defined objectives (Paulson, Paulson & Meyer, 1991).

An effective portfolio is *continuous and ongoing*, providing both formative and summative opportunities to monitor the student's progress towards achieving essential outcomes. It is *multidimensional*, inasmuch as it reflects a wide variety of artifacts and processes mirroring various aspects of a student's learning process. Furthermore, it contributes to *self-reflection*, including ways for students to reflect about their own thinking processes and meta-cognitive introspection as they analyze their own comprehension, reflect on their approaches to problem-solving and decision-making, and observe their growing understanding of subjects and skills. Approaches to portfolio development vary, yet all of the major literature emphasizes the following characteristics: They clearly reflect defined learner outcomes regarding the curriculum that students are expected to study. They focus on the student's learning experiences as well as his acquisition of crucial knowledge, skills, and attitudes. They contain samples of work that stretch over a period (e.g. a semester), rather than single time points. And they contain a variety of work samples and evaluations of that work by the student and teacher (George, 1995).

The underlying rationale behind portfolio assessments can be summed up by educational philosopher John Dewey's (1916) observation, that children and adults alike learn most naturally when they have a problem solving experience with relevant, real-life issues. Cognitive psychologists and constructivist learning theorists have come to view learning as a construction of the individual, not something that is merely absorbed from teachers and books. Thus problem solving, student reflection, and their appropriate portrayal or documentation, are increasingly receiving primary attention (Jones, 1994).

Portfolio types

There are different types of portfolios, each serving one or more specific purposes as part of an assessment program. The most common types are the following.

The *documentation portfolio* (also known as the "working" portfolio) involves a collection of work over time showing growth and improvement reflecting the student's learning of defined objectives. It may include everything from brainstorming activities to drafts to finished products and becomes meaningful when specific items are selected out to focus on particular educational experiences or goals. The *process portfolio* documents aspects or phases of the learning process and is particularly useful in documenting the student's overall learning process. It can show how students incorporate specific knowledge or skills and advance towards basic and advanced mastery. It also depicts the student's reflection on his learning process, including the use of journals and related forms of meta-cognitive processing. The *showcase portfolio* is mainly used for a summative evaluation of a student's mastery of crucial curriculum results and includes his very best work. This type of portfolio is mostly compatible with the development of visual artwork and consequently often used at art schools.

Portfolio development

The development of a portfolio can be divided into three phases (Burke, 1994):

1. Organizing & planning (consists of exploring and deciding on essential questions at the beginning of the process. Students can thus fully understand the purpose of the portfolio and its status as a means of evaluating their own progress)
2. Collection (of meaningful artifacts and products, reflecting the student's educational experiences and goals. Based on the intent and purposes of the portfolio decisions must be made about the context and contents)
3. Reflection (demonstrates the student's meta-cognitive contemplations on the learning process and his monitoring of his increasing understanding of essential knowledge and skills)

Grading portfolios

In order to evaluate a portfolio, teachers must have more than one scoring strategy to evaluate the student's progress. Criteria for the assessment of a finished portfolio can include:

- Thoughtfulness (evidence of the student's self-reflection)
- Growth and development in relationship to crucial curriculum objectives

- Understanding and application of focal processes
- Comprehensiveness, accuracy, and adequacy of products and processes presented
- Heterogeneity of entries and artifacts

It is especially important for both teachers and students to work together to prioritize those criteria that will be used as a basis for assessing and evaluating the student's performance (Burke, 1994). The fundamental framework for an evaluation, however, remains the feedback discussion that can equally include characteristics of an academic discussion or elements of a traditional oral exam.

Portfolio-assessment at the MILAK

Assessing students' performance with portfolios is a relatively new approach at the Swiss Military Academy. Hubert Annen, Ph.D., head of military psychology and military pedagogy studies, started in 2006 to evaluate the academic performance of prospective professional officers using process portfolios. Teaching topics in both the three-year Bachelor Study Course as well as the Diploma Study Course and Military School I (both one-year) include psychological theories (e.g. motivation); the young adult; military education; normal and traumatic stress; authority, power and obedience; and psychological warfare. The main goal of the lectures consist on the one hand in imparting descriptive and procedural knowledge (i.e. the "what" and the "how"), on the other hand, however, the objective is to develop a culture of human-oriented[1] leadership and raise the students' awareness for both the potential as well as the limitations of their future role as educators.

Possible pitfalls that arise as a result of living and working in a total institution such as the army and wielding a considerable amount of power over young and susceptible young adults need to be addressed. Not simply in order to prevent or deal with one's own shortcomings, but also to identify, confront and handle subordinates' excesses and aberrations in military training and education. The aforementioned advantages of portfolio-assessment (i.e. responsibility, self-reflection, self-initiative and self-criticism) all contribute to raising the students' awareness and developing their resources in order to transfer these insights into practice.

In the specific case of our study courses, possible contents of a portfolio may include documentation of a lesson, current newspaper clippings relating to topics

1 A term coined by Brigadier Prof. Dr. Rudolf Steiger, former director of the MILAK. Core of this concept is the conviction that subordinates in their capacity as human beings play a key role in achieving group-defined goals.

of the lecture, summaries, memos, graphic reduction and visualization of data, learning journals, practical examples and own experiences, and commentaries by peers or colleagues. The criteria used to evaluate the portfolio are fourfold and include:

1. Arrangement (e.g. table of contents, lay-out, composition)
2. Artifacts & products (e.g. completeness, relevancy, uniqueness, scope of the selection)
3. Reflection (e.g. originality, substance, structure)
4. Learning process (e.g. development, autonomy, strengths and weaknesses)

Experiences and preliminary data

Our experiences concerning portfolio-assessments, as succinct as they may be, have shown us that students more willingly accept the grades they receive for their work, as the feedback discussion allows them to better comprehend the rationale behind the grade. Incidentally, on a more anecdotal level, the acceptance this new kind of evaluation tool received by the students was, although not negative, far from exuberant, as all realized that it implicated an additional effort during the semester. Furthermore, due to the relatively unspecified and open nature of a portfolio, the seemingly unclear guidelines unsettled many students. Even after having finished their portfolios and receiving their grades, there weren't many rave reviews. However, the same students who were critical of portfolio-assessment right up to the feedback discussion stated about 6 months later, that only then had they realized the beneficial effect the ongoing and self-directed engagement with pedagogic and psychological issues had had on them. They felt that their developing of a portfolio had had a far more sustainable and long-term effect than they had expected initially. It is thus obvious, that the acceptance and effectiveness of portfolio-assessment needs to be evaluated not merely at the end of semester, or after receiving the grades, but after allowing the subject taught to develop a long-term effect.

To date, the only available feedback data consists of one Bachelor Study Course (N=12) from 2006. The data suggests that the majority became more aware of their learning process and reflected more on their learning behavior, and almost all students considered their grades as fair and reasonable and the feedback discussion as equitable and pleasant. The majority also got to know their standard of knowledge better and regarded the tool as beneficial in transferring their knowledge into practice. These advantages, however, are gained not only at the expense of an above-average mental and emotional presence of the students; they also involve an increased effort on part of the evaluating teacher. The evaluation of a portfolio requires a great deal more time than a classic pencil and paper test. This downside is nevertheless worth the effort, as the prospect of achieving sustainable and practice-effective results in a crucial and important field of education such as officers training fully makes up for this extra time.

Although by no means statistically significant, these findings leave us confident of the potential value of portfolio-assessments in the education and training of career officers. With an ever-growing database, we will soon be able to make reliable and valid statements as to the effectiveness of this uncommon, yet valuable evaluation tool.

References

Annen, H. (2002). Action Research as a Method for Scientific Thinking and Acting in Military Pedagogy. In H. Florian (Ed.), *Military Pedagogy: An International Survey* (pp. 227-244). Frankfurt a.M; Berlin; Bern; Bruxelles; New York; Oxford; Wien: Peter Lang.

Annen, H. (2007). Military Pedagogy and Ethics: Current Challenges in Switzerland. In J. Toiskallio (Ed.), *Ethical Education in the Military* (pp. 183-195). Helsinki: Edita Prima Oy.

Burke, K. (Ed.) (1994). *The Mindful School: The Portfolio Connection*. Palatine, IL: IRI/Skylight Publishing.

Dewey, J. (1916). *Democracy and Education. An Introduction to the Philosophy of Education* (1966 Edn.). New York, NY: Free Press.

Feldman, R. (2003). *Epistemology*. Boston, MA: Prentice Hall.

George, P. (1995). *What Is Portfolio Assessment Really and How Can I Use It in My Classroom?* Gainesville, FL: Teacher Education Resources.

Jones, J. (1994). Overcoming Resistance to Self-Direction in Adult Learning. In R. Hiemstra & R. Brockett (Eds.), *New Directions for Adult and Continuing Education* (pp. 23-29). San Francisco, CA: Jossey-Bass Publisher.

Paulson, F., Paulson, P. & Meyer, C. (1991). What Makes a Portfolio a Portfolio? *Educational Leadership, 48* (5), 60-63.

Antti-Tuomas Pulkka

Strategies and Challenges in Officer's Training and Education

Introduction

The purpose of this study is to analyse the Strategy of the National Defence University (NDU) from a viewpoint of teachership and to critically illuminate and address points of interest. The Departments of the NDU and their teachers should be informed, guided and inspired by the Strategy of the University. The question is therefore: *What does the Strategy mean to an individual teacher?*

For interpretation, a text can be studied by using various approaches of which a deep approach, seeking meaning behind mere words and contents, is most permissible in this context. For this purpose, the text of the Strategy of the NDU was analysed qualitatively to re-arrange it and to critically view the challenges that face those immediately dealing with implementation. The reduced data was displayed in compressed categories labelled according to principal discoveries of the text and these categories are explained and discussed in the light of their projected practical consequences.

Findings implicate that it seems that the demands of the Strategy are formidable. The roles of the teacher and teaching are seen to be strongly related and even integrated with scientific research. A teacher will be facing multiple challenges when trying to balance between the expectations of learners and the requirements of the department and collegial processes of research.

Teacher's view of the Strategy

The Department of Education is part of the National Defence University and its teachers should be informed, guided and inspired by the Strategy of the University. Teachership includes didactic choices and planning with the aim of ever developing the contents and delivery of courses and lectures. The teacher at the NDU (at least at the Department of Education) can quite freely decide from a wide array of methods and practices and sometimes even the actual substance for his or her teaching. Some of these choices are premeditated, some quick everyday decisions and some are even unconscious. How should the premises for these choices be influenced by the Strategy?

The Strategy of the National Defence University is such a high-level document, that it does not present or limit specific procedures, but it should give general and philosophical guidelines to steer the teachers to a desirable course. (Amin & Cohendet 2004, 5-7) The Strategy should – at least partly – act as the constellations did for sailors in ancient times; they guided on a rough course and after that it was a question of seamanship whether they made right interpretations of other signs at the right moment in a rapidly changing environment, as experts of modern times will have to do (Amin & Cohendet 2004, 52; Engeström 1992,17).

As a document of a high purpose and meaningful intentions, the Strategy of the National Defence University provides descriptions of these "constellations" for users to "draw their own maps", as teachers should do as development-oriented experts (Engeström 1992, 11, 14-16, 26; 2004, 152-153). As a product of human activity the text is to be seen as a collection of symbols expressing layers of meaning[1]. For interpretation, a text can be studied by using various approaches of which a deep approach, seeking meaning behind mere words and contents, is most permissible in this context.

For understanding the activities of teachers and departments it is vital not only to present the words of the strategy, but also to recognise the variety of possible interpretations of its key points. (Vanhala et al. 2002, 93) The challenges of implementation can only be justifiably discussed when they are seen in the right context i.e. the starting point: What is the situation now in the departments? How much do these prerequisites vary? What are the distinctive features of the departments, which cause these differences? To answer these questions (though ambiguous they are!) is to answer the question: *What does the Strategy mean to an individual teacher?*

Qualitative analysis of the text of the strategy

To interpret (and so to understand) the Strategy in a credible way it is analysed qualitatively. This brief analysis is made by partly applying the general views of Miles & Huberman (1994) so that the analysis is seen as consisting of three concurrent flows of activity: reduce data, display data and draw conclusions. The data is displayed in compressed categories labelled according to the principal discover-

1 Although constructivism is very righteously criticised because of its most radical forms, it is seen here that constructivism as a theory concerning scientific knowledge presents *knowledge as social construct.* (cf. Matthews (2000, 63) according to Puolimatka (2000, 38-39)) For more criticism and discussion on possibilities and limitations of constructivism cf. example Gergen 1995 and Lincoln & Guban 2000.

ies of the text and these categories are then explained and discussed in the light of their projected practical consequences.

Two assumptions were made to direct the reduction of the utterances. First, it is assumed that because of its iterated and socially (militarily!) constructed nature, the Strategy should hold its primary message in some of the leading sentences in the first parts (Introduction and the first chapters). Second, it is assumed that the increased frequency of a particular utterance in the text will signal something of its importance. Qualitative analysis is sometimes criticized of subjectivity, but explanatory power can be increased in this study by making the process somewhat transparent and therefore the major steps are described in detail.

The examination was started by searching the first part of the text for utterances about either specific expressions or radical demands for the basics of the college.[2] The following quotations of utmost importance arise from the textual construction of the leading parts (author's italics):

"Military sciences is a broad concept and a multifaceted wholeness, that is bound together by the studies of wars, crises and relevant security threats and endeavours to prevent them." (Introduction)

"According to valid regulations the National Defence College plans and facilitates the teaching in the Officers Training Program and military scientific research is conducted at the College." (Chapter 1.1)[3]

"The National Defence College holds the status of a university. This previously meditated policy is also supported by scientific research, the nature of the degrees and the present administrative composition [of the college]." (Chapter 1.1)

"Upon 2016 the National Defence College will be a highly appreciated university – both domestically and internationally – concentrating on defence and security matters." (Chapter 2)

At this phase the criteria for choosing utterances was complemented and reconsidered (Tuomi & Sarajärvi 2002, 110).[4] Preceding sentences were synthesized so that further reduction was directed by the ideation of the NDU as a "university facilitating broad-based and multifaceted teaching and conducting scientific

---

2 Miles & Huberman (1994, 69) call this phase *generating pattern codes*, where "researcher is looking for threads that tie together bits of data" during initial fieldwork and it is seen here that authors experience and intentions naturally guide this part of the process.

3 According to new policy, the official name of National Defence College is nowadays National Defence University but as the Strategy is dated earlier than this policy, the former name is displayed in the quotations.

4 Tuomi and Sarajärvi discuss on the dynamic and self-adjusting nature of qualitative analysis so that the process itself is under constant change when new things are learned from the material. Miles & Huberman (1994, 69-70) also describe pattern coding as a dynamic action where researcher must avoid "getting locked too quickly".

research". This ideation is actually also the traditional definition of a modern university.

By using this hereby specified criteria, the following essential utterances (among others) were chosen for further processing:

"The societal influence of the National Defence College focuses on its part on the development of defence and security policy." (Chapter 3.1)

"As a part of the Finnish university system the National Defence College is acting on an equal level of science and education with other universities and it is serving its purpose commissioned by the National Defence and Security Policy." (Chapter 1.2)

"At the National Defence College teachers conduct independent research and teaching is founded on this research." (Chapter 1.2)

"The status of research is strengthened and its productivity is improved by increasing research personnel and funding." (Chapter 4.1)

"Degrees in Military Sciences are developed to be analogous to competences needed in the future." (Chapter 4.2)

"Information technology is widely utilized in research, teaching, studying and administration." (Chapter 4.4)

"Research activities aim to elaborate knowledge of the physical area of action competence." (Chapter 4.5)

"The high quality of the NDC equals the excellence of research, teaching, studying and administration. The core of the quality is innovativeness, criticality and the ethic responsibility of the students, teachers and researchers." (Chapter 4.8)

The order of the samples above is of their relative importance in the thematic of this examination and altogether there were 79 hits as the unit of the analysis was a sentence. The findings were characterized among other things by repeating utterances of multifarious nature of the NDU`s subjects, working across disciplines and research and development guiding subjects and teaching. According to the interaction of utterances (cf. Miles & Huberman 1994, 69-72), four categories were formulated and labelled as following:

- Scientific research and development activities (31 hits)
- Teaching/instruction as independent activity (14 hits)
- Wide-ranging and cross-discipline (15 hits)
- Research as an integral part of teaching (19 hits).

The utterances related to teaching or interacting factors are those demanding change from the traditional role of the teacher of NDU. It is my experience, after six years in the NDU, that mostly – though not thoroughly – new teachers are recruited not for their virtues in the academic context but for their expertise on a definite professional area, their positive attitude and drive and willingness. It is not assumed that this is just so at all the departments. After all this is an interpretative examination from a very individual point of view.

A word from the departments

Challenges are illustrated by results to a questionnaire that was administered to the heads of departments. They were asked to categorise their personnel on different roles on teaching and researching. Teachers were to be placed in two categories whether they conduct actual scientific research (in traditional meaning) or if they are involved in more informal, yet still research-oriented developmental work. Also the heads of the departments were asked if they predict change in these figures in the near future. The statistics are reported in Table 1. The results indicate that only in one department (dep. of Strategic & Defense studies) the roles of the teachers are consistent with the idea of the strategy about teacher's role. The other departments reported that most of their teachers have nothing to do with scientific research or small-scale research and development work. The organisational structure of the departments and the stated or perceived role of teachers do not support the vision at the moment, although some positive predicaments can be made.

Department	Teachers total number	Doing research	Doing Research and development work	Predicted change
Education	23	7	6	no answer / cannot predict
Management and leadership	14	6	no answer	Fig 2 to increase to 9
Tactics and operations	32	6	14	Fig 2 to increase to 8
Military technology	15	6	1	Fig 2 to increase to 8
Strategic and defence studies	10	10	10	no change predicted
Military history	6	6	2	Fig 1 & 2 increase to 7

Table 1. Teacher's involvement in research in the departments of NDU

Discussion

The Strategy of the NDU presents major requirements to teachers. It should be accepted at every level that the teaching profession is changing from the traditional teacher-centred sharing of professional knowledge to a more self-regulated, reflective guide and tutor of evermore demanding learners. The teacher will have to be able to critically observe his/her own practices, to be conscious of his/her true goals and to renew his/her schemes (Kulkki-Nieminen, 2001, 70-71). A balance will have to be found between the traditional role and the revised nature of teaching in NDU as the Strategy also underlines that professional knowledge derived from experience has to be utilized also.

There are two aspects that will have to be recognized, also on other levels than rhetoric:

1) The teacher must do research on his/her field of expertise even at a modest magnitude at the beginning so that the substance is based on scientific validation and deep understanding.
2) The teacher will be spending much more time in the "preparation" of a course than before if the first aspect is taken seriously – the repetition of an old slide show should not be an option (Poikela 2001a, 24).

Retaining and repetitious teaching does not fulfil the requirements of high-quality academic teaching that the Strategy is demanding. Should there be no research and predictions and critical discussion as a natural part of teaching, our students would only learn things valid and relevant for the moment (or yesterday), and that is also not an option. The retaining role of the teacher will prevail, if the teachers are not required, encouraged and resourced to study their profession and do research on their discipline. (Poikela 2001b, 107-108) Altogether, teachers will need a *new kind of expertise*, traditionally conceptualized *expert as an excellent individual* will not suffice but teachers will have to be future-oriented and learning-oriented, their knowledge cannot be only accrued from routines and previous achievements. (Engeström 1992, 11, 14-16, 26; 2004, 145-153) The demands faced are therefore high, and it remains to be seen how our organization will cope and what measures are needed in different departments so that the change is both possible and tolerable.

The strategy can be interpreted in various ways depending on the point of view and it is therefore a matter of the utmost importance that open and continuous discussions are held when implementing the strategy to make the official interpretation transparent and acceptable. Should the definite measures taken and the superiors' interpretations be controversial to those of the subordinates, the lack of discussions may induce some implicit and latent frustrations that might appear as unwanted features in the process of teaching.

References

Amin, A. & Cohendet, P. (2004). *Architectures of Knowledge – firms, capabilities and communities*. England: Oxford University Press.

Engeström, Y. (1992). Interactive Expertise – Studies in Distributed Working Intelligence. *Research bulletin 83*, Department of Education, University of Helsinki. Helsinki University Press.

Engeström, Y. (2004). The new generation of expertise – Seven thesis. In H. Rainbird, A. Fuller & A. Munro (Eds.). *Workplace learning in context*. Englanti: Routledge.

Gergen, K.J., (1995). Social Construction and the Educational Process. In L.P. Steffe. & J. Gale (Eds.) *Constructivism in Education*. USA: Lawrence Erlbaum Associates, Inc., Publishers.

Kulkki-Nieminen, A. (2001). *Tukeeko ainelaitos opettajuuden kehittymistä? Teoksessa Tutkiminen on oppimista – ja oppiminen tutkimista*. [Do departments support the development of teachership? To research is to Learn – to Learn is to Research.] Tampere: Tampereen Yliopistopaino.

Lincoln, Y.S. & Guba E.G. (2000). Paradigmatic controversies, contradictions, and emerging confluences. In K.N. Denzin & Y.S. Lincoln *Handbook of Qualitative Research*. 2nd Edition. USA: Sage Publications, Inc.

Miles, M.B. & Huberman A.M. (1994) *Qualitative Data Analysis*. 2nd edition. London: Sage.

Poikela, E. (2001a). *Tarvitaanko yliopistossa lehtoreita? In Tutkiminen on oppimista – ja oppiminen tutkimista*. [Does University need Scholars? To research is to Learn – to Learn is to Research] Tampere: Tampereen Yliopistopaino.

Poikela, E. (2001b). *Ongelmaperustainen oppiminen yliopistossa. In Tutkiminen on oppimista – ja oppiminen tutkimista*. [Problem-Based learning in University. To research is to Learn – to Learn is to Research] Tampere: Tampereen Yliopistopaino.

Tuomi, J. & Sarajärvi, A. (2002). *Laadullinen tutkimus ja sisällönanalyysi*. [Qualitative inquiry and content analysis] Helsinki: Tammi.

Vanhala, S., Laukkanen, M. & Koskinen, A. (2002). *Liiketoiminta ja johtaminen*. [Business and management] Keuruu: Otavan kirjapaino Oy.

Luiza Kraft

Eurocompetences for E.U. standards in the military education system – The case of foreign langugage teaching in "Carol Ist" National Defense University, Bucharest, Romania

The paper is based on the results of a survey originating in a Needs Analysis questionnaire applied in March 2007 on foreign language teachers of "Carol 1st" National Defense University of Bucharest, Romania. The data collected point to the new competences required from the teachers of English working in the Romanian defense-related education system (i.e., scientific, digital, critical, linguistic, and cultural literacy), and identifies the new roles that the language educator must play nowadays: facilitator and guide to the learners, mediator, researcher and designer of complex learning scenarios, collaborator and evaluator. One of the core conclusions of the survey is that teachers, alongside learners, should be trained to meet the special needs of interaction and co-operation with 'otherness'.

Introduction: A new language teacher for new teaching roles

The goal of this paper is to profile the *'new' teacher* of foreign languages *in a European perspective* and, in particular, to establish which competences are required from him/her – in accordance with the provisions of the Common European Framework for Reference – to be able to prepare our military students for effective work in multinational groups or commands, or on the theaters of operations worldwide.

Following the Maastricht Treaty (Art. 126 and 127), there has been undoubtedly increasing awareness of a European dimension in education and, for foreign language teachers, this concept is inevitably linked with the belief that foreign language study without a context of cross-cultural communication is sterile and risks being limited to the knowledge of purely formal data. Learning a language involves elements which are linguistic as well as cultural and, although foreign languages play an extremely important role in enlarging the relations among peoples, teaching still needs to be linked much more closely with the cultural aspect, which is the only way to achieve awareness and appreciation of 'otherness' as a source of intellectual growth.

Raising linguistic and cultural awareness in military students is a must and also a many faceted problem. The cornerstone of education is, as stated for SOCRATES, "to create the bases necessary to reach the professional competences required by a teaching system aimed at the learning of communicative competence in an intercultural dimension and interaction among the different cultural realities" (Nalesso, 2003).

As we have moved into the 21st century, the role of the modern language teacher in Europe has substantially changed and in a variety of ways. These changes are driven by different social, economic, political and cultural trends. Among these trends we can mention:

1) increasing internationalization of the economic activity;
2) increased cross-cultural and cross-linguistic interaction through tourism and virtual communication;
3) increased mobility of labor;
4) increased demand for modern languages as mediums for everyday communication instead of, or as well as, mediums for high culture;
5) political decisions about the need for generalized linguistic competence in at least one modern language and perhaps two[1], and in consequence:
   - multi-level teaching of target modern languages;
   - teaching of modern languages lower down the age range;
6) increased public interest in all professions including teaching.

Language education as a factor in social development

In the Council of Europe's work over more than thirty years, there has been a constant and developing assertion of the importance of the social role of language education.

The *Common European Framework* states as one of its objectives "to promote methods of modern language teaching which will strengthen independence of thought, judgment and action, combined with social skills and responsibility"[2].

As a result of the needs analysis questionnaires meant to identify short and long term teacher training priorities and to suggest programs for instruction, the teacher trainers working in the Romanian MELT[3] observed a low status for language teachers in many Language Centers of the Romanian Ministry of Defense and considered that better understanding and communication of the role of language learning in individual and social development was a key issue for their work; this

1 the first NATO language is English, and the second is French
2 Common European Framework for Reference, 2001, p. 4.
3 Military English Language Teaching

led to the need for principled work and hence to designing and running projects concerning the status of language teachers and exploring their future role.

At this point we were able to state the important social role of language educators and to inquire what the over-riding aims of language education are.

Therefore, the resulting aims include:

- the development of European citizenship, with an educated European understanding several languages, able to study and travel in many countries, knowledgeable about, and with respect for, many different nationalities and national cultures;
- the conviction that knowing different languages is a powerful factor in intellectual development, encouraging open-mindedness and flexibility, contributing to the development of other skills;
- the commitment to life-long language learning. The aim of school education, at any level, should be to train students to become good language learners, capable of acquiring the particular languages as they meet the need for them;
- the idea that language study offers opportunities to acquire independence and autonomy as learners, that it can be learnt in ways which encourage co-operation and other social values (Heyworth, 2003).

What kind of modern language teacher?

This change in role is more demanding in many ways than the traditional role of the modern language teacher. It therefore requires qualities in addition to those, which were typical of the good modern language teacher of the past.

According to the answers given by the participants in the survey, among those traditional professional qualities that will still be required are[4]:

- *proficiency in the target language;*
- *knowledge of, and a positive attitude towards, the target culture;*
- *knowledge about language as a system and social construct, with particular reference to the target language;*
- knowledge of appropriate assessment techniques.

4 *Living together in Europe in the 21st century: the challenge of plurilingual and multicultural communication and dialogue.* Proceedings of the third colloquy of the European Centre for Modern Languages (Graz, Austria, 9-11 December 1998).

Certain personal qualities also resulted to be traditionally esteemed in the good modern language teacher, which need not be detailed here.

In addition to these, the changing demands are considered to require new professional qualities. The most important five chosen in the questionnaire are:

- *much more sophisticated and effective pedagogical skills*, motivated by the need to teach across a wider spectrum of abilities, across a wider age-range to students, whose motivation to learn modern languages (or indeed to learn any school subject) which is something that will have to be aroused rather than assumed; the skills will include awareness of new developments such as the applications of information technology to language teaching;
- *much greater autonomy and self-direction in professional development, meaning* that teachers will have to be more pro-active in self-development, otherwise they will inevitably eventually be found wanting as attainment targets are changed and (probably) raised;
- *particularly for in-service teachers, greater ability to "bootstrap" their self-development through exploiting their own professional practice,* by being able to reflect on that practice in a structured way, and possibly in partnership with colleagues;
- *sensitivity to pluricultural and multicultural issues and a willingness to work in* international *programs*;
- ability to exploit and build on school and teaching experience.

Teachers and Learners: new roles and competences

New competences

This part of the paper is an overview of the new skills expected of teachers in a drastically changing learning environment like the military education system in Romania and in any NATO country, especially due to the use of new media, the status of Romania as a full-fledged NATO member and EU member starting Jan. 2007.

Anthony Fitzpatrick (2003) lists five types of new literacies that teachers need to understand and master alongside learners. They are: scientific, digital, critical, linguistic, and cultural literacy.

Scientific literacy relates to the ability to think scientifically in a world, which is increasingly shaped by science and technology. This kind of literacy requires an understanding of scientific concepts as well as an ability to apply a scientific perspective.

Digital literacy relates to the ability to use ICT adequately and apply them in a principled way to the subject matter at hand. For the language teacher, it refers in particular to Web literacy, i.e. the ability to make use of the World Wide Web for language research; to the use of linguistic tools and standard programs for exercises and testing.

Critical literacy implies the ability to evaluate the credibility, usefulness and reliability of any given sources of information. It also encompasses skills in sifting and identifying the relevant and important in the flood of information, which threatens to engulf the unprepared.

Linguistic literacy in this context refers to the ability to recognize different genres as they develop, to track developments in language use and usage and to adapt materials (authentic or not) for teaching.

Cultural literacy relates to observing and recording changes in the society or societies of the target language together with implications for language teaching. Such changes may be of a general nature leading to convergence between own, native culture and the target culture, or to changes particular to the target culture.

Furthermore, in order to function adequately in the world of the new media, teachers need to acquire and master a whole range of new skills ranging from the technical, to the organizational and the conceptual.

Teachers need to become completely computer-literate and have the confidence to use the available technology adequately. They should be able to cope with the most common problems arising from the use of computers very much in the way that average car drivers can cope with commonly occurring problems with their motor vehicles, i.e. no specialist knowledge of the machine, but knowing what to do when routine breakdowns occur. This task would probably become easier for the teachers of tomorrow who are the learners of today.

Given the momentum gathered in the nature of out-of-school and inside-school innovations, new organizational and pedagogic models are called for, including ICT for teacher education using a learning by doing and reflective approach. The innovative potential of languages going online must be fully grasped, where teachers can build and sustain language communities; dismantle them when they have exhausted their function; and link minds and hearts in order to negotiate everyday concerns or even complex issues. Language is a social activity for which you need real partners for communication and teachers increasingly need to diversify the partners with whom their learners practise the target language.

New conceptual skills are required of teachers when one considers that there is a quantum leap in moving from well-tried controllable media like the textbook with its supplementary materials, to the more open, inquiring approach when exploiting the new media to the full. Teachers must move to a role in which they are designing learning experiences and planning encounters for / with their learners with the target language environment in mind, or – more preferably – at hand.

The importance of intercultural communicative competence should also be highlighted and education in this respect focused on.

Teachers, alongside learners, now need to be able to meet the special needs of interaction and co-operation with 'otherness'. Aleksandrowicz-Pedich et al. (2003) analyzed a questionnaire that had been distributed in English in ten countries, and in French in seven countries, regarding the views of teachers of English and French on intercultural communicative competence in language teaching. Overall they found some divergent abilities in teachers across the spectrum. It is significant that while teachers are aware of the importance of the issue, they agree that the move from ideological acceptance to practical solutions ought to be facilitated by training and availability of proper class materials.

New roles and attitudes

Teachers are required to take on new roles and come prepared with the pertinent attitudes to the classroom. As in contexts of autonomous learning (Holec & Huttunen, 1997), today's teacher has more and more increasingly to function as facilitator and guide to the learners. Some other new and important teacher roles are those of mediator, researcher and designer of complex learning scenarios, collaborator and evaluator.

The teacher as facilitator and guide. As facilitators, teachers must in many ways know more than they would as directive givers of information. Facilitators must be aware of a variety of materials available for improving students' language skills. In such a context where bottom-up processes are adopted, this focus on choice and independent use of materials by students under the teacher's guidance has been identified in terms of a pedagogy of resources, in parallel with other pedagogies like the *pedagogy of time*, the *pedagogy of choice* and the *pedagogy of cooperation* (Balbi, 1993).

As facilitators, teachers have to be flexible, responding to the needs that students have, and not dependent on just what has been set up ahead of time by a curriculum developer and their idea of who will be in the classroom. Teacher education is a key element to success in this more flexible language classroom so that teachers will have the ability to use and to recommend multimedia and other resources effectively. In fact, teachers must not only know about and understand the functions of different media available in a media-rich environment, they should also know when best to deploy them.

In the joint construction of projects with their learners (a pedagogy of cooperation among learners, and of teachers with learners), teachers need to guide learners in the use of word-processing, graphics and presentation programs. The integration of audio-visual elements will bring home to learners the fact that the foreign language environment of the target language is as vibrant and multi-faceted as the society in which they live.

The presentation of material in a creative manner also aids the development of cognitive processes. In order to help learners to enlarge their mental and social abilities, to learn how to learn, teachers themselves are a major asset in guiding learners and at the same time practising alongside them, skills like hypothetical thinking, analysis of point of view, and the application of different symbol systems including the graphical, the musical, and even the mathematical as is now possible through the integration of multi-media environments in the language learning process.

The teacher as mediator. The role of mediator is not new for language teachers as it has always been their task to act as intermediary between two cultures while they introduce learners to new linguistic and cultural concepts. However, the immediacy offered by the new media forefronts this role and gives it a new weight. Within the relatively safe confines of traditional textbooks, teachers could introduce relevant aspects of the target language and culture in small, manageable chunks. Access to the 'real world' of the target culture and, at times, confrontation with it, requires new strategies and approaches that need to be learned and practised.

A more realistic departure point is that which takes into account the multilingual environment that learners live in and bring with them into the classroom. What is required, first and foremost, is a move to a multilingual perspective, a pluralist conception of language experience, which takes into account individuals' varying levels of contact with and competence in a range of different languages, and supports the development of their plurilingual competence.

The teacher as researcher. To keep abreast of developments in relevant fields of knowledge and walks of life, particularly in the countries of the target language, teachers need to know how and where they can access information for their own and for their learners' use. Furthermore, reflective practice and action research are becoming part and parcel of the daily lives of teachers. Teachers need to keep up to date with knowledge generated in the field of modern languages and applied linguistics, not only for the sake of their learners, but also for their own professional development that is often taking the shape of 'reflection-on-action'. Action research involves a self-reflective, systematic and critical approach to enquiry by teachers in order to identify areas of difficulty, to embark on a period of study, possibly in collaboration with others, and to bring about informed changes in practice as a result.

Teachers-as-researchers should be able to adjust continually to varying demands made on their professional knowledge. These requirements vary from, for example the level of learner competence they are asked to deal with, to the increasing responsibility placed on them to participate in the administration and evaluation of the educational institution.

The teacher as designer of (complex) learning scenarios. In order to orchestrate successful learning scenarios, teachers need to learn how to put together tasks and materials to guide their learners to successful execution and conclusion of their projects. Unlike work with conventional teaching materials (textbook, workbook, audio and video materials), which have been graded, pre-assembled and collated in a chronological order, the design of learning scenarios nowadays is much more complex, requiring higher order skills involving researching and evaluating source materials, setting overall aims and objectives and breaking down tasks into meaningful and manageable sequences.

The complexity of learning environments also demands that teachers be able to switch between a variety of roles such as encourager to learners in establishing learning objectives; as task-setter in providing learning tasks; as guide throughout the various stages of the learning process; as presenter when providing information and as evaluator both during and at the end of a task. The flexibility demanded of teachers is related not only to the changing nature of current methodologies, or to the changing type of learners they face from year to year, or from group to group, but also to the range of roles they have to perform within the same course and with the same group of learners.

The teacher as collaborator (with other teachers). The investment of time and effort required in this new learning environment implies a sharing of responsibilities and tasks among teaching staff. Collaboration with colleagues will lighten the burden and make the efforts more fruitful and rewarding.

New management patterns must emerge to ensure fair distribution of workloads, and revised job descriptions will be necessary to share and co-ordinate the tasks in hand.

The current practice of language subjects that compete for time and space on the timetable may need to be altered as already mentioned above. One of the implications of this is surely related to the need for much greater collaboration among language teachers.

The teacher as orchestrator (technology, learners, curriculum). Teachers will need to develop fairly sophisticated management skills in order to be able to provide a healthy balance between the different elements, which make up the new learning environments. Mastery and confidence in the use of technology needs to be applied to the learning inclinations and abilities of individual learners whilst covering the prescribed syllabus or curriculum, which are often set by outside authorities for schools.

When the teacher is conversant with the learning styles of the students, and is able to bring the learner style and the learning method and tools into synchronization, then the symphony orchestrated by the teacher as conductor with the learners as performers will surely be more in tune. This is quite a sophisticated task as the teacher needs to be able to identify with precision the learning styles of learners; to choose from and apply with efficacy the relevant learning techniques, tasks and materials; and thus to spark off and successfully sustain the learning process, often within set frameworks which are less than ideal or conducive to collaborative learning.

The teacher as evaluator and self-evaluator. The first evaluation task for a teacher is probably that of selecting materials, methods, and other means for the learners to work with. Furthermore, evaluation of both the learning process and the product, e.g. student level of competence acquired, call for a radical revision of current models of evaluation like examinations.

When Frank Heyworth (2003) talks about the new tools for language learning such as the *Common European Framewor*k, the *European Language Portfolio* and the *European Language Passport*, he forefronts the relevance of 'process' in learning. In this way a humanistic approach to learning and to evaluation can be sustained. There is no doubt that new teacher education practices will also reflect this trend of empowerment through introspection and self-evaluation and assessment. Teacher self-evaluation also includes self-assessment of the teacher training and development programs, self-reflection on the impact of personal professional growth on the learners' progress, as well as on the teaching team's development.

In this way there is a greater chance for teachers to succeed in their role of raising the learners' awareness of learning, and empowering them to carry out self-assessment.

Conclusion: The need for education and professionalisation

As results from the research work done in preparation of this paper, the cornerstone of teacher formation is that of practice-oriented training. In the arguments outlined above, in fact, we have identified various practical skills needed which lead us to the conclusion that a theoretical knowledge base becomes inadequate unless it is accompanied by the various strands of practical methodology. Furthermore, what we are arguing for here is the need for the student-teacher to experience learning (and teaching) in the classroom and not only or mostly in the university lecture hall. It is imperative for teachers to acquire first-hand experience of the methods used in classrooms.

In relation to this is the principle of reflective practice. The reflective model of teacher preparation has gained prominence in recent years. It goes very well alongside the principles of learner autonomy, social and personal development and intercultural competence, which have now become accepted as key areas of language learning.

Given the revised dimension of teacher as learner, the teacher's role includes an on-going process of reflection in order to evaluate what happened in the classroom and why, and as a result to have the confidence to experiment and try out new ideas alongside the learners themselves[5].

Some key aspects of teacher professionalism performed as a result also include: reflective partnerships between teachers, peer-mentoring and school-based enquiry. It goes without saying that these practices necessitate teamwork and collegiality and help break down teacher isolation and stagnation.

It is only in this way that Romanian teachers of English can successfully face the challenges of the new NATO and EU membership.

References

Aleksandrowicz-Pedich, L. (2003) ECML project 1.2.3: Intercultural communicative competence in teacher education, "The views of teachers of English and French on intercultural communicative competence in language teaching", In: I. Lázár, (Ed.) *Incorporating Intercultural Communicative Competence in Language Teacher Education*, Strasbourg, European Centre for Modern Languages / Council of Europe.

Balbi, Rita. (1993). Autonomous learning in the classroom. In: *Workshop 13A Language and culture awareness in language learning / teaching (L2 and L1) for the development of learner autonomy (age 11-18)*, Strasbourg: Council of Europe.

Fitzpatrick, T. (2003) Teacher Training, In: Anthony Fitzpatrick (ed.) *Project 1.3.1A 'ICT in VOLL'*. Strasbourg: European Centre for Modern Languages / Council of Europe.

5 The European Commission, 1995. *White Paper on Education and Training. Teaching and Learning. Towards the Learning Society*. Luxembourg: Office for Official Publications of the European Community

Heyworth, F. (2003) Introduction – a new paradigm for language education. In: Frank Heyworth (ed.) *Facing the Future. Language Educators across Europ*e. Strasbourg: European Centre for Modern Languages / Council of Europe.

Holec, H. & Huttunen, I. (1997) *Learner Autonomy in Modern Language*s. Strasbourg: Council of Europe.

Nalesso, Marilena. (2003) Teacher Competences for a Linguistic and Cultural Education towards Mobility in Europe. In: Bärbel Fink (Ed.), *Which diversification and how can it be achieved?* Graz: The European Center for Modern Languages Publications.

The Authors

Annen, Hubert, Ph.D., Head of Military Psychology and Military Pedagogy and Dean of the Teaching and Research Branch at the Military Academy at ETH Zurich. Head of the Assessment Center for career officer candidates. Research in military pedagogy, evaluation and validation of assessment and selection processes for military leaders, in motivational aspects in military context and in personality and organizational attitudes.
hubert.annen@milak.ethz.ch

Hall, David Ian, D.Phil., (Oxon.), Senior Lecturer, Defence Studies Department, King's College London, at the Joint Services Command and Staff College, UK Defence Academy. Research in modern military history, and modern German history.
dhall.jscsc@da.mod.uk

Jung, Hermann, Ph.D., ret. 2002. Since 1962 Artillery officer in the Austrian Armed Forces, teacher in fire direction and artillery observation in the Artillery School, 1972 assigned to the Austrian Army Command as expert for Instruction technology and Curriculum development, following studies in adult-education leadership development at the University Klagenfurt, Austria. Further studies at the University of Vienna, Austria. Before retirement head of the defence pedagogy and civic education subdivision of the Training and Education Section of Ministry of Defence, Austria. Member of the International Association of Military Pedagogy. Publisher and Co/editor of the publication Series on Military Pedagogy, Military Sciences & Security Policy, Peter Lang. Focus on military pedagogy and political/cultural education.
gabihe@etel.cc

Kasher, Asa, Laura Schwarz-Kipp Professor Emeritus of Professional Ethics and Philosophy of Practice at Tel Aviv University and co-author of the first IDF code of ethics.
asakasher@hotmail.com

Kraft, Luiza, PhD, Associate Professor at the Foreign Language Department, the Command and Staff Faculty, of "Carol I" National Defense University - Bucharest, Romania. Head of the English Department and teacher trainer. Research in organizational communication, intercultural education and English language teaching methodology.
luiza_kraft@yahoo.com

Libel, Tamir, B.A Tel- Aviv university, M.A Bar Ilan University where he is finishing a PhD Dissertation, both in Political Studies. His dissertation compares the changes in command and staff colleges, advanced war-fighting institutions (such as the UK Higher Command and Staff Course) and War/ Defense/ Security colleges in the US, UK, Australia and Israel between 1991- 2003. He has published and presented papers on topics including military education, Israeli military doctrine and Israel's military history. He is affiliated with the Begin Sadat Center for Strategic Studies at Bar- Ilan University and is member in the International Association of Military Pedagogy, Israel Political Science Association and the Association for Israel Studies.
tamirlibel2@gmail.com

Mäkinen Juha, Acting Professor, Ph.D., LtCol., Department of Leadership and Military Pedagogy at the Finnish National Defence University. Research in military pedagogy, social and knowledge sciences.
juha.makinen@mil.fi.

Malos, Gavril, Ph.D, Commandor (r) professor, „Carol I" National Defence University of Romania, Department of Education and Human Resources. Research in project management for Military Pedagogy and Human Resources.
gabriel_radu2000@yahoo.com

Mardar, Sorina-Mihaela, Lecturer, PhD, "Carol I" National Defence University of Romania, Department of Education and Human Resources. Research in continuing professional education, European dimension of Romanian legislation concerning adult professional education, interactive training methods used in adult education.
sorinamardar@gmail.com

Müller, Antonín, RNDr. CSc., Head of Curriculum Department University of Defence in Brno, Czech Republic, member of University Pedagogical Board. In a long-time horizon he is interested in providing pedagogical process at universities

with special focus on military ones, structure of pedagogical staff and condition for their professional growth and its modeling.
antonin.muller@unob.cz

Plesanu, Toma, Colonel professor, PhD, „Carol I" National Defense University of Romania, Department of Education and Human Resources. Research in project management in Education and Human Resources.
tomaplesanu@yahoo.co.uk

Pulkka, Antti-Tuomas, Captain, doctoral student (University of Helsinki), Course Director at the Continuing Training Division of the Finnish National Defence University. Research in military pedagogy, ethics and motivation in learning.
antti-tuomas.pulkka@mil.fi

Rossi, Philip J., S.J., Professor of Ethics and Systematic Theology, Marquette University, Milwaukee, Wisconsin, USA. Research on just war theory, on religion and human rights in global culture, and on the political and social philosophy of Immanuel Kant and its contemporary import.
philip.rossi@marquette.edu

Royl, Wolfgang, Prof. ret, Dr. phil., University Federal Armed Forces, Munich, Department of Education.
W.Royl@unibw-muenchen.de or W.Royl@t-online.de

Saliger, Radomír, Mgr. Ing. Ph.D. Head of social scientists at the Deaprtment of Social Sciences and Law, University of Defence in Brno. He is a member of Czech Pedagogical Society, European Security Association, and Czech Association of Pedagogical Research. He publishes internationally, specializing in the field of education and training of military professionals, in social communication and rhetoric. He also participates in sociological research for the military. He is in the University of Defence senate and is a member of Pedagogical Board of the university.
Saliger@seznam.cz

Simpson, Harold, MBE MA, Lieutenant Colonel. Head of UK Defence Instructor Training and Development at Defence Centre of Training Support, RAF Halton, Aylesbury, UK. Research into military pedagogy, development techniques and the use of military education a mechanism for initiating reform and Nation Building in post conflict societies.
209simps@armymail.mod.uk

Syme-Taylor, Victoria, Director of External Relations, Defence Studies Department, Joint Services Command and Staff College UK. The Defence Studies Department is part of the War Studies Group King's College London. Research and policy advice on military education reform. Military Education as part of democratic reform.
VSyme-Taylor.jscsc@defenceacademy.mod.uk

Talerud, Bo, Ph,D. Associate Professor, Head of Education and Leadership within the department of Leadership at the Swedish National Defence College. Research in military education and leadership, civil-military relations in international operations and ethical aspects on armed forces. Member of the IAMP board.
bo.talerud@fhs.se

Studies for Military Pedagogy, Military Science & Security Policy

Editors:
Dr. H. Jung
Prof. Dr. W. Royl

Band 1 Edmund A. van Trotsenburg: Militärpädagogik. 1989.

Band 2 Hermann Jung / Heinz Florian: Grundlagen der Militärpädagogik. Eine Anleitung zu pädagogisch verantwortetem Handeln. 1994.

Band 3 Franz Kernic: Demokratie und Wehrsystem. Aufsätze zum Verhältnis von Gesellschaft, politischem System und Heer in Österreich. 1997.

Band 4 Edwin R. Micewski: Grenzen der Gewalt – Grenzen der Gewaltlosigkeit. Zur Begründung der Gewaltproblematik im Kontext philosophischer Ethik und politischer Philosophie. 1998.

Band 5 Rudolf Egger / Heinz Florian (Hrsg.): Pädagogische Professionalisierung im Bundesheer. Dokumentation und Reflexion des PädAk-Sonderstudienganges Wehrpädagogisches Management. 1999.

Band 6 Franz Kernic / Harald Haas: Warriors for Peace. A Sociological Study on the Austrian Experience of UN Peacekeeping. 1999.

Band 7 Franz Kernic / Jean M. Callaghan / Philippe Manigart: Public Opinion on European Security and Defense. A Survey of European Trends and Public Attitudes Toward CFSP and ESDP. 2002.

Band 8 Heinz Florian (ed.): Military Pedagogy – An International Survey. 2002.

Band 9 Edwin R. Micewski / Hubert Annen (eds.): Military Ethics in Professional Military Education – Revisited. 2005.

Band 10 Hubert Annen / Wolfgang Royl (eds.): Military Pedagogy in Progress. 2007.

Band 11 Hubert Annen / Wolfgang Royl (eds.): Educational Challenges Regarding Military Action. 2010.

www.peterlang.de

Peter Lang · Internationaler Verlag der Wissenschaften

Hubert Annen / Wolfgang Royl (eds.)

Military Pedagogy in Progress

Frankfurt am Main, Berlin, Bern, Bruxelles, New York, Oxford, Wien, 2007.
291 pp., num. tab. and graph.
Studies for Military Pedagogy, Military Science and Security Policy.
Edited by H. Jung and W. Royl. Vol. 10
ISBN 978-3-631-56072-3 · pb. € 56,80*

Armed forces need to prepare their military personnel for a global mission profile which requires conflict awareness and the acceptance of personal responsibility along with the ability to successfully cope with the military challenges of their mission. The aim of this publication is to demonstrate the progress in military pedagogy as it is evolving within the various national institutions for the education and training of officers and recruits. This will be helpful in illustrating common topics which, already at present, allow the questioning and exploring of educational ways of thinking as they determine national peculiarities in the training and education of soldiers.

Contents: Advanced strategies in Military Pedagogy · The military and its society · Educational aspects of military leadership · Application areas of Military Pedagogy

Frankfurt am Main · Berlin · Bern · Bruxelles · New York · Oxford · Wien
Distribution: Verlag Peter Lang AG
Moosstr. 1, CH-2542 Pieterlen
Telefax 0041(0)32/3761727

*The €-price includes German tax rate
Prices are subject to change without notice

Homepage http://www.peterlang.de

Lightning Source UK Ltd.
Milton Keynes UK
UKHW021257170722
405967UK00003B/13

9 783631 589410